T0247626

Dinosaurs at the Dinner Party

HOW AN ECCENTRIC GROUP OF VICTORIANS DISCOVERED PREHISTORIC CREATURES AND ACCIDENTALLY UPENDED THE WORLD

Edward Dolnick

SCRIBNER

NEW YORK   LONDON   TORONTO   SYDNEY   NEW DELHI

Scribner
An Imprint of Simon & Schuster, LLC
1230 Avenue of the Americas
New York, NY 10020

First Scribner hardcover edition August 2024

SCRIBNER and design are trademarks of Simon & Schuster, LLC

Simon & Schuster: Celebrating 100 Years of Publishing in 2024

For information about special discounts for bulk purchases, please contact
Simon & Schuster Special Sales at 1-866-506-1949 or business@simonandschuster.com.

The Simon & Schuster Speakers Bureau can bring authors to
your live event. For more information or to book an event,
contact the Simon & Schuster Speakers Bureau at 1-866-248-3049
or visit our website at www.simonspeakers.com.

Interior design by Kyle Kabel

Manufactured in the United States of America

1   3   5   7   9   10   8   6   4   2

Library of Congress Cataloging-in-Publication Data

Names: Dolnick, Edward, author.
Title: Dinosaurs at the dinner party : how an eccentric group of Victorians
discovered prehistoric creatures and accidentally upended the world / Edward Dolnick.
Description: First Scribner hardcover edition. | New York : Scribner,
[2024] | Includes bibliographical references and index.
Identifiers: LCCN 2024003627 (print) | LCCN 2024003628 (ebook) |
ISBN 9781982199616 (hardcover) | ISBN 9781982199630 (ebook)
Subjects: LCSH: Paleontology—Great Britain—History—19th century. | Paleontologists—
Great Britain—History—19th century. | Dinosaurs—Great Britain—History.
Classification: LCC QE705.G7 D65 2024 (print) | LCC QE705.G7 (ebook) |
DDC 567.90941/09034—dc23/eng/20240409
LC record available at https://lccn.loc.gov/2024003627
LC ebook record available at https://lccn.loc.gov/2024003628

ISBN 978-1-9821-9961-6
ISBN 978-1-9821-9963-0 (ebook)

For Lynn, and Sam and Ben

How does the once unthinkable become not only thinkable but self-evident?

—Lorraine Park, historian (in a 2019 interview)

Contents

CONTENTS

A Shriek in the Night

In the early decades of the 1800s, the world was a safe and cozy place—the natural world, that is, the living world of trees and flowers and darting fish and soaring birds and leaping deer. "It is a happy world," wrote William Paley in 1802, in one of the era's most influential books. "The air, the earth, the water, teem with delighted existence." Everywhere he turned his gaze, Paley went on, "myriads of happy beings crowd upon my view."

Paley was a deeply knowledgeable student of natural history, not a daydreamer wandering barefoot through the woods. He was both a philosopher and a clergyman, and in England in the years around 1800, few thinkers were as well-known or well-regarded.

The more closely you looked at nature, Paley argued, the more proof you found that the whole system had been flawlessly designed. It was true on the largest scale: the whale's mighty heart was a feat of oversized engineering—a child could crawl through a whale's aorta. And on the smallest scale: the human eye and ear made use of muscles so tiny they were miracles in miniature, "and yet," Paley marveled, "the grandest and most precious of our faculties, sight and hearing, depend upon their health and action."

It would have been easy to imagine a less imposing world. Animals might not bound like deer or gallop like horses but drag themselves along the ground with a single stubby hoof. Flying creatures might not soar and glide but tumble through the air like broken kites.

Reality was more inspiring. Every creature was an ingenious living machine, and all those creatures interacted in perfect and eternal balance. God's wisdom and foresight shone out everywhere you looked. Consider which animals were rare and which were common. "Birds of prey seldom produce more than two eggs," Paley noted approvingly, so that the skies weren't thick with dive-bombing raiders. Harmless sparrows or ducks, on the other hand, might have a dozen eggs in the nest. In the sea, we find "a million herrings for a single shark."

How had such smooth-running complexity come to be? Not by chance, certainly. To look at the world as it hummed along was to marvel at the God who had designed it all.

Better yet, the Creator was as kind as he was wise. The happiness that we see all around us was not a fortunate surprise; it was the point of the whole operation.

"The world was made with a benevolent design," Paley declared. The humblest creatures displayed their glee as unmistakably as fifth graders bursting out of school at the end of the day. Even newborn flies flitting about, with their "gratuitous activity, their continual change of place without use or purpose, testify their joy and exultation." Fish are "so happy that they know not what to do with themselves."

Everyone ate it up, scholars and laymen alike. At Cambridge University, for instance, Paley was mandatory reading for decades. His influence on the university curriculum, one historian writes, "was second only to that of Isaac Newton."

No other domain offered a refuge anywhere near as enticing as the natural world. Certainly the human landscape did not. In precisely the era that Paley described so rosily, mayhem and violence reigned on a global scale. "The first world war began in 1793," the historian John Ray observed, "and lasted until the Battle of Waterloo in 1815."

Battles in this era were easy to romanticize—poets wrote of cavalry charges and wooden ships with billowing sails—but a truer picture would have shown vast landscapes turned into slaughterhouses, with hundreds of thousands of soldiers blasting away at close range. The death toll came to three million soldiers and one million civilians. "The lesser nations of Europe were at best pawns," in one historian's summary, "and at worst the chessboard."

The trouble began in France, where a baying mob cut off the king's head, and later that year the queen's, and then Napoleon's dreams of glory left half of Europe soaked in blood. In England, a much-feared invasion never came, but life was hard and precarious.

Food was scarce, taxes were high (to pay for the wars), and an age-old way of life, based on agriculture and small-scale craftwork, had come under siege. Workers with sledgehammers broke into factories and smashed the machines that had taken their jobs.

Crime was commonplace, and the law was ruthless. Two hundred crimes, from poaching a rabbit to robbing a coach, carried the death penalty. In 1808 reformers achieved one small victory—picking pockets was no longer punishable by death.

This was an era of rapid technological progress—factories and rail-roads were the emblems of the age—but also the era of *Oliver Twist* and *David Copperfield*, of "Please, sir, I want some more" and workhouses and debtors' prisons. The books were fiction, but the desperation and hardship came straight from life. Children as young as four worked long hours amidst whirling machines that threatened at any moment to grab a dangling sleeve or a careless finger.

It is hard to overstate the gulf between the commotion and chaos of this new world and the quiet, soothing scenes that Paley described so lovingly. For generations, the rhythms of life had scarcely changed. Workdays had always been governed by the rising and setting of the sun and the changes of the seasons.

No more. Now industrialists and inventors had harnessed the power of coal and steam and iron, and sprawling factories ran day and night. Thousands of workers at a time spilled into the mills, worked twelve-hour shifts, and staggered home again.

Even for the well-off, life was not all waltzes and dinner parties. London, the biggest city in the world, was filthy and foul smelling. The air was black and sooty; the streets were a mire of mud and horse dung that sucked at pedestrians' shoes and grew ever more slick and treacherous as countless hooves and wheels splashed their way along; the Thames was thick with human waste and dotted with animal corpses. In one historian's summary, "Rome in the first century AD was a far cleaner place than London eighteen hundred years later."

Nor had medicine advanced in all those centuries. In a world without antibiotics and with hardly any understanding of hygiene, death was a

This drawing of death patrolling the Thames ran in *Punch* magazine with the title *The Silent Highwayman*. The drawing appeared in 1858, when it had just been demonstrated that cholera was a waterborne illness. Until that revelation, cholera had seemed to strike at random.

lottery and disease ran unchecked. "King Cholera" stirred special fear. No one knew where it came from or how it spread. Cholera swooped down in waves and could kill a thousand people at a time. A person might be healthy at breakfast and dead before dinner.

Faced with such chaos, one impulse was to turn away in search of refuge. The inanimate world offered one sort of haven. It was orderly, above all, with the stars shining and the planets making their stately rounds. But perhaps it was a bit *too* orderly, elegant and precise but as hard to fall in love with as a diagram in a geometry text.

The living world, in contrast, was not only perfectly arranged but inviting, too. If the astronomers' realm was analytical and austere—a Bach minuet would have provided a fitting soundtrack—the naturalists' domain was cheery and vibrant—imagine Vivaldi's "Spring" concerto.

Restrict your view as the naturalists did, and the world's hurly-burly seemed far away. Life purred along as cozily as in a cottage plunked down in a country garden.

Then, in 1802, the very year of Paley's "happy world," something ominous shrieked in the night outside those cottage windows.

On an ordinary day in New England, in the town of South Hadley, Massachusetts, a twelve-year-old farm boy set to work plowing his father's fields. The young man carried an imposing name—Pliny Moody—but nothing in his life had ever marked him as special or seemed likely to.

On this particular day, though, Moody's plow turned up something that caught his eye. When he climbed down from his horse and cleared away a bit of mud in the turned-up field, he found himself staring at a long row of footprints. There were about a dozen prints in all, each the size of a dinner plate, with three toes on each foot. The tracks stretched across a slab of mud that had long ago turned to solid stone.

The tracks on that New England farm, it would turn out, had been made two hundred million years before, by a dinosaur. In 1802 no one had ever heard of dinosaurs. These were, as far as anyone knew, the first dinosaur tracks ever found.

*   *   *

That find was as strange and unexpected as any discovery in human history. A series of similar discoveries followed, in rapid succession and across the globe. The finds were giant bones and enormous footprints in stone and, soon, immense skeletons. No living creatures looked like this. *Whose bones were these?*

Today every child knows the answer. But in the early decades of the 1800s, no one had any inkling that there had ever been such creatures as dinosaurs. The word *dinosaur* would not be coined until 1842.

That seems awfully late in the day, especially considering how much we now take dinosaurs for granted. Today every natural history museum in the world boasts an enormous dinosaur skeleton that scrapes the ceiling and pokes its head out into the hall. Every kindergarten class has a plastic dinosaur or two near the shelves with the crayons and the blunt-nosed scissors. Every toddler has pajamas with cartoon dinosaurs or a bin stuffed with toy dinosaurs.

But a time traveler from 1800 would look at those toys and relics in bewilderment. Shakespeare, who imagined everything, never imagined a world ruled by house-sized beasts and where human beings had never set foot. The world's most farseeing thinkers had never seen that far. No such possibility had ever crossed the mind of Leonardo da Vinci or Galileo or Isaac Newton or Benjamin Franklin.

Until around 1800, in fact, scientists and laymen alike had taken for granted that the world had always looked much as it still did, with the dogs and daffodils and oaks and horses that we all know. The idea that Earth had a history—that it had once been home to vastly different forms of life— had scarcely occurred to anyone. (The phrase *natural history* dates back to ancient times, but there had never been any history in "natural history.") And only a few bold thinkers had imagined that the world was not a mere few thousand years old but hundreds of thousands of years old or millions.

The standard picture was far easier to grasp and far more assuring. All the world's creatures—and human beings, too—had recently been set in place by God, like figurines in a diorama.

God had crafted that scene with exquisite care, and he had lavished special care on humankind, his favorites, fashioned in his image. Now, suddenly, that cozy picture no longer fit. But what was the alternative?

Outside of myths and fairy tales, no one had ever dreamed that creatures like three-toed giants had once rambled across the land. No one ever imagined that, for eons and eons, legions of flying, slithering, lumbering creatures had ruled the world, and that human beings had played no role whatsoever in the tale.

And if anyone *had* somehow conjured up such scenes, they would never have gone farther and imagined that all those animals could have vanished. A person or an animal could die, of course, but no one had ever dreamed that a *species* could die.

The idea of extinction is so familiar today—toddlers know that pandas are endangered—that it is worth emphasizing that it is a relatively new idea. It had almost no place in nineteenth-century thinking. God had fashioned the world, and he was a perfect designer whose works were eternal, not an artist who ran out of inspiration and flung his misfires into a wastebasket.

The thought of permanent, arbitrary disappearance from the ladder of life was as unnerving for our forebears as would be the news, for us, that whole groups of people going about their everyday lives—crowding into a commuter train in the morning or picnicking in the park on a Saturday—might suddenly vanish into nothingness.

The dinosaur discoveries came in a rush in the early 1800s, partly because the Industrial Revolution brought a frenzy of digging of all sorts. As armies of workmen tore up the ground with picks and shovels, their canals and mines and tunnels and quarries offered never-before-seen peeks beneath Earth's surface.

Fossils, which had turned up occasionally through the centuries (and never been understood), were now encountered almost routinely. Stones from quarries were used for road mending, one geologist noted cheerily to a lecture audience in 1836, so "You will ... probably crush beneath your carriage-wheels the remains of creatures which, had you lived a hundred thousand years ago, might have turned the tables, and crushed you."

The bones were old, but the way of thinking that asked *How can we learn what these bones really were?* was new. Gigantic bones had turned up through the centuries, but people had been more inclined to incorporate them into age-old myths than to truly examine them.

By the 1800s, as science began to come into its own, better answers seemed called for. But the dinosaur story stands apart from other scientific mysteries even so. It is an unusual story, and it is best approached in an unusual way.

Most scientific mysteries have a straightforward structure—someone points out something odd, and people gather clues and suggest theories in an attempt to explain it. Detective stories work the same way—a woman walking her dog sees a body with its throat cut, half-hidden behind a bush. Police cordon off the crime scene and start investigating. Near the body is a muddy footprint, in a trash bin is an empty wallet, on a security camera is a fuzzy silhouette.

The dinosaur story was different, in large part because the sleuths did not begin with a victim—a body—that they could examine and ponder. Instead, most often, they had only a few bones or teeth, and their task was to *imagine* a body from those scanty hints.

It was a bit like trying to solve a maddeningly difficult jigsaw puzzle where a great many pieces had been lost, and pieces from different puzzles had been flung together, and no one had ever seen the picture they were trying to assemble.

Worse still, many puzzle pieces had been picked up, admired for their handsome appearance, and then put to one side because no one recognized that they had any special significance. (A clue is not a clue until someone sees a mystery.) Momentarily intrigued by the sight of a fearsome tooth or a giant rib, people through the centuries had installed the remarkable object in a conspicuous place and soon ceased thinking about it.

Today we encounter these puzzles from the past in their complete, solved form, with all the signs of hard work hidden from view. The resolution of the tale is so familiar—*Dragonlike creatures roamed an ancient world*—that it is easy to forget how hard-won those successes were.

We will spend most of this book in the first half of the nineteenth century, when the puzzle pieces fell into place. But in order to give the heroes of our saga their due, we will also circle back to earlier centuries, when talented and earnest thinkers veered badly off course. We will dive into those all-too-human tales, too, where would-be sleuths eagerly gathered up blue puzzle pieces in the happy belief that they would form the sky, only to learn much later that the picture was an indoor scene, and there was no sky, and no two blue pieces ever fit together.

There were good reasons for those bad guesses, and it will be worth our while to take a few excursions upstream in time, to learn how our forebears saw the world.

Our story is set mostly in England, because that's where the first dinosaurs turned up. But we will make forays into France and America (Thomas Jefferson will play a walk-on role). We'll venture briefly into China and Greece, too, where huge bones that turned up in ancient times gave rise to myths about dragons and griffins and the race of man-devouring giants called cyclopses.

It's our good fortune that the scientists we will come to know best were odd and outsized themselves, like the bones they unearthed. Their ranks include an impoverished, barely educated woman from England's southern coast who began collecting fossils to earn a few pennies from wealthy tourists and eventually became the greatest fossil finder of them all; a wildly eccentric Oxford geologist whose home looked as if Noah's ark had run aground at the front door and spilled its contents into his rooms; and England's best-known (and most hated) scientist, whose talent for identifying extinct animals was surpassed only by his gift for making enemies.

The tale begins around 1800 and reaches its climax on New Year's Eve 1853. Those bookend dates are key. These were early days for science—the word *scientist* did not exist until 1834—and natural history in particular was an unsettled muddle where nearly every question was up for debate.

The dinosaur discoveries were news flashes in a world unprepared to make sense of them. Suddenly it seemed that the familiar world had been built atop a vanished world, or perhaps a series of vanished worlds, that had been filled with gigantic marauding creatures.

The cozy cottage had become a haunted house.

CHAPTER I

"Dragons in Their Slime"

For our ancestors in the early 1800s, the discovery of bones and footprints was thrilling, bewildering news. This was not just another scientific discovery, like the sighting of a new moon around a distant planet. This was proof of *life* where no one had ever imagined it.

Think of modern-day astronomers who believe that we on Earth are not alone. For decades they have turned their telescopes to the heavens, searching for signals in an endless void. So far, nothing.

Scientists and fossil hunters in the first half of the 1800s were the frock-coated counterparts of those present-day searchers. With two crucial differences—they looked back in time, not out in space, and they *did* detect signs of life.

Not subtle signs either, like odd patterns of static picked up by a computer. Here were teeth like daggers and ribs like rafters.

Poets, scientists, and ordinary men and women looked at the dinosaur discoveries and shuddered and marveled. Tennyson (using an archaic word for "tore") imagined a bygone world that featured "Dragons of the prime, / That tare each other in their slime."

Perhaps Tennyson and his peers would have been less astonished if they had anticipated a world that teemed with huge, violent beasts. But

This mayhem-filled drawing is from an 1851 book called
A History of All Nations, about history and prehistory. Violent clashes
of prehistoric creatures were a favorite theme of early illustrators.

the nineteenth century's aliens turned up out of the blue, as we have seen, and the public was blindsided in a way that people in today's world—who have known about the search for ET for decades—could never be.

For a moment, try to imagine a modern counterpart. Think what it might be like if people had never dreamed of life anywhere but on Earth. And then picture that one night a spaceship materialized a few dozen feet above Fifth Avenue and proceeded to make a slow and stately tour of Manhattan.

One of the great hazards in trying to picture the past is forgetting that our forebears didn't know how their story ended. We read about the Great Depression or the rise of Nazism knowing how it all played out. It's hard to bear in mind that no one in the 1930s had that privilege. But we flatten out history's drama—we miss out on people's fears and hopes and illusions and expectations—when we bring our present-day knowledge with us on our ventures into past ages.

In the case of the dinosaur discoveries, it wasn't simply that no one knew how the story ended. More important, no one even knew what to make of how the story *began*.

That puts the dinosaur story into unusual company. Every once in a great while, people going about their ordinary lives have looked up and seen something they never imagined. A ship with towering masts and billowing sails materialized on the horizon, for instance, in waters that had never known a vessel bigger than a canoe. Or a stranger turned up in a valley so remote that its inhabitants had thought themselves alone in the world.

Of all such first encounters, none ever topped the moment when humans first stumbled on bones, footprints, and other evidence that dinosaurs had once roamed the earth.

What was it like to see what no one had ever seen before?

The dinosaur story took on its modern shape around 1800, but it *almost* began far earlier, back in 1677. In that year, workmen digging in a quarry about twenty miles from Oxford University found a massive bone. They brought it to Robert Plot, a much-admired naturalist and the first curator of Oxford's Ashmolean Museum. Plot was seldom at a loss—he was always referred to as "the learned Dr. Plot"—but now he stared, perplexed. This was surely "a real bone, now petrified," he wrote, but *whose* bone?

The bone was broken at one end, but the break was neat, and the intact piece looked to Plot "exactly" like the lower part of a thigh bone. Except that *this* femur was huge, at more than two feet around and about twenty pounds in weight. This was, Plot grumbled, too big to make sense. "It will be hard to find an Animal proportionable to it," he wrote, "both horses and oxen falling much short of it."

Plot ventured a guess. "It must have been the Bone of some Elephant," brought to Britain more than a thousand years before when the Romans had invaded.

That was a good idea, but Plot admitted that he had his doubts. Other giant bones had been found in England, he pointed out, and if those were from elephants, too, why was it that no one had ever found any of "those great Tusks with which they are armed"?

And what of other giant bones that had recently been dug up in a churchyard near Bristol? Were *those* from elephants, too? Plot confessed his bewilderment: "Now how Elephants should come to be buried in Churches, is a Question not easily answered."

Then came the final blow to the elephant theory. "There happily [i.e., fortunately] came to Oxford while I was writing this," Plot recalled, "a living Elephant." This was not an instance of the circus coming to town. Plot meant that his museum had received the skeleton of an elephant from the present day rather than one that had lived a thousand years before.

Plot rushed to compare his mystery bone and other giant bones in the museum's collection with the elephant's bones. Nothing matched! So horses were out, and oxen were out, and now elephants were out, too. What was left?

Plot spelled out the only remaining possibility. "Notwithstanding their extravagant Magnitude, they must have been the Bones of Men or Women."

By way of support for this eye-catching claim, Plot went on to provide a pages-long list of human giants throughout history. Some had been described by Greek and Roman authors in antiquity, and some were more recent. The physician to the queen of Hungary had declared, about a century before, that "there dwelt a Person within five Miles of him ten Foot high."

Plot cited other cases. France had boasted a giant of its own, at about the same time that Hungary's giant was roaming the countryside. The "Giant of Bordeaux" was so tall, reliable witnesses had reported, that "a Man of an ordinary Stature might go upright between his legs when he did Stride." For Plot, as for many others, biblical accounts carried the most weight of all. "Goliath for certain was nine Foot nine Inches high."

Case closed! The broken thigh bone came from a giant human being.

From our vantage point, this seems silly. But Plot was not a silly man. He was open-minded and methodical, and he had gathered all the evidence he could find. But he was, like all of us, a creature of his own era. Which meant, in his case, that he could not imagine *other* eras and *other* creatures and a world before humans.

Wrong guesses like Plot's continued long after the learned doctor's death. Eighty-six years after Plot first pondered it, the giant femur received its first official scientific name. The name reflects a striking fact—almost a century after Plot had decided that so enormous a bone could only have come from a human giant, science continued to endorse the same view.

In 1763 an English physician and naturalist named Richard Brookes compiled a six-volume encyclopedia of natural history. Brookes did not have access to the actual bone that Plot had studied, which had been lost at some point along the way. But Plot had drawn it carefully and recorded all its measurements, and that was nearly as good. Brookes reprinted Plot's original drawing exactly.

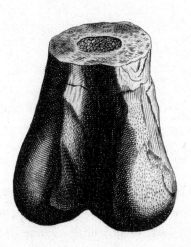

This was indeed a petrified relic from a human giant, Brookes concluded, but it had nothing to do with thigh bones. What we had here was plainly a pair of enormous testicles from a bygone human giant. Brookes bestowed an imposing Latin name on the fossil, *Scrotum humanum*. (A

French naturalist named Jean-Baptiste Robinet took matters a step farther, in the same misguided direction, by identifying signs of musculature and a bit of urethra.)

The bone would eventually be properly identified, but not until 1824. It would take that long to come up with an explanation that would have struck Plot and his successors as far more outlandish than a world replete with human giants.

CHAPTER 2

The Girl Who Lived

When Pliny Moody spotted footprints in his father's field, there was no precedent he could turn to for guidance. None that anyone he knew had ever heard of, at any rate. But if there was no known precedent in history, there was one—nearly—in literature.

One of the most unsettling scenes in fiction occurs in one of the first novels ever written in English. Daniel Defoe published *Robinson Crusoe* in 1719, just over three hundred years ago. It is to this day one of the most reprinted books in world literature, by some tallies second only to the Bible.

The story shifts gears in one startling scene. Crusoe has been marooned for fifteen years, condemned to live and die alone on an empty island. Walking near the water's edge at about noon one day, he looks down and discovers a bare footprint—distinct, definitely not his, and unmistakably human—in the sand.

Terrified, Crusoe whirls around to see who could have made it, but there's no one there. Crusoe flees to his makeshift camp, looking over his shoulder as he runs, mistaking every bush and tree for a person.

Days later, he convinces himself that he must have imagined the whole scene. He forces himself to return for another look. And he sees the footprint again, still intact, still far different from the mark of his own foot, still impossible to explain away.

That print on a sandy shore carried an unnerving message—*You are not alone*. The three-toed prints on a Massachusetts farm carried almost the same message—*Creatures beyond your imagining passed this way before you.*

Pliny Moody stumbled on his dinosaur footprints by accident in 1802, and he never knew just what he had found. The most accomplished fossil finder of them all came along just a decade later, and she knew precisely what she had achieved.

Mary Anning was a poor, uneducated young woman who grew up in Lyme Regis, a small town on the English Channel. Her life was hard from the outset. Dickens would have blushed to invent so bleak a story. Anning, who was born in 1799, was one of nine or perhaps ten children—the *perhaps* is telltale, for countless lives in nineteenth-century England went almost completely unrecorded, and no one is certain of the size of the Anning family. Only Mary and an older brother survived childhood. The family lived in a small house next to the town jail.

Mary was the second child in the family to bear that name; she was named for an older sister who had burned to death when her clothes caught fire on Christmas Day, or thereabouts, in 1798. She had been playing near the fireplace and had been "left by the mother for about five minutes," the local newspaper reported. Five minutes was all it took. The family had buried another child, an infant named Henry, earlier the same year.

The second Mary was born five months after her sister's death. On an August day the following summer, when Mary was a year old, a group of traveling performers came to town. A storm blew in. With the rain pelting down, everyone took shelter as best they could, in hopes that the horseback tricks and riding and jumping would soon resume.

A woman from the neighborhood had charge of Mary for the day. Clutching the baby to her body, she gathered up two other children and retreated under an elm tree. Then came a jolt of lightning and an enormous clap of thunder. The tree shattered. Mary's minder fell to the ground, dead, still holding Mary tight in her arms. Both of the other children were killed at once. Mary showed no signs of life either, but someone carried her home. Her parents put her in a warm bath, and soon afterward she opened her eyes.

Lyme Regis was a resort town known for its dramatic setting and its abundant fossils. (It achieved brief fame in modern times as the setting for *The French Lieutenant's Woman*.) Richard Anning, Mary's father, was a carpenter and cabinet maker who brought in a bit of extra money by selling "curiosities" pried out of the cliffs to well-to-do tourists. (Jane Austen was one of those summertime visitors. She even had a run-in with Richard, when she felt he'd asked too high a price for repairing a broken lid on a box.)

How did an out-of-the-way place like Lyme Regis come to play an outsized role in our story? Because a fluke, or, really, a series of flukes, combined to make the small town one of the few places in the world where a fossil finder could hope to earn a living.

It was happenstance that had made the unstable cliffs along the Dorset coast rich in fossils. (Geologists would eventually sort out how cliffs that rose into the sky came to be studded with the bones of creatures that lived in the sea. But that would come later.) It was happenstance that brought tourists with thick wallets and a taste for souvenirs to Lyme Regis. And it was happenstance that made limestone a good material for building, which meant that quarrymen had dug into the cliffs for many years and, in the process, had unearthed a steady supply of bones and shells turned to stone.

Lyme Regis is a tourist destination today, with ice cream shops, beachside cafés, and restaurants outfitted with oars, buoys, and other bits of nautical debris. The tasteful street lighting is decorated with ironwork in the shape of ammonites, which are coiled fossils that look a bit like nautilus shells. But until the mid-1700s, no one would have dreamed of visiting Lyme Regis, or any other seaside town, for a holiday.

No one strolled by the water's edge or played in the waves. No resorts catered to tourists. Everyone knew that the coast and its waters were the natural haunt of rough, weather-beaten characters like smugglers and fishermen. The gentry looked, shuddered, and kept their distance.

An ammonite from the Lyme Regis Museum

That disdain had deep roots. For ancient Greeks, one historian notes, the sea was a place of terror and dark secrets, "a closing-over element filled with voracious monsters." Even in the Middle Ages, the sea conjured up images not of sun and sand but of whirlpools and whales and shipwrecks, and pirates and raiders.

The sight of water extending as far as the eye can see was frightful, not soothing or inspiring. Even nature lovers agreed. As late as the 1850s, Thoreau would warn that "the ocean is a wilderness reaching around the globe, wilder than a Bengal jungle and fuller of monsters."

The tide turned, so to speak, in the mid-1700s, and medical quackery played a large role in the shift. The vogue began when an English physician named Richard Russell revived an ancient doctrine that salt water had healing powers. Immensely influential, Russell convinced the fashionable world that God had created the sea as "a Kind of common Defense against the Corruption and Putrefaction of Bodies."

Dr. Richard Russell, whose *Dissertation on the Use of Sea Water in the Diseases of the Glands* inspired a fad for holidays by the sea

The wealthy had long sought to revive their health at spas and mineral springs, and now Russell advised his readers to take advantage of the sea, as well. Russell moved his own practice to Brighton, a seaside town, and built a house—the largest in town—that doubled as a kind of sanitarium.

Well-off travelers who had long flocked to spa towns like Bath now began to swarm to coastal spots like Brighton, Margate, and Lyme Regis. A dip in the sea (and even a drink *from* the sea) was recommended for ailments from epilepsy to rheumatism to nervous troubles. (One of the best-known fossil collectors, Thomas Hawkins, first visited Lyme

Regis because his doctor had advised him to try sea bathing as a cure for deafness.)

Then, in the early 1800s, fear of Napoleon helped the fad along. With Europe in disarray, wealthy English families began to opt for holidays closer to home. In the meantime, improved roads and the rise of the railroad made travel easier. That dealt the middle class into the game, too, and soon fishing villages all along England's coast transformed themselves into tourist resorts.

England had already pioneered a host of industrial advances. Now it added a different sort of invention to its tally—the beach holiday.

Fossils were a family business for the Annings, with everyone recruited to hunt for relics and then to scrub and polish the good ones. Richard set up a wooden display table just outside the front door. Mary began helping out at age five or six.

This was a competitive business. The Annings and other collectors scanned the beaches and cliffs for fossils and then scrambled to peddle their trinkets. Exactly what fossils were, neither buyers nor sellers knew. They were amulets with magical healing powers (for the superstitious) or knickknacks (for those who wanted a souvenir to set on the mantelpiece) or natural wonders (for the scientifically minded).

From the start, Mary had a knack for spotting a bit of treasure peeking out an inch or two from a crumbling cliff or a newly eroded rock slide. But fossil hunting was hard, dangerous work that involved clambering over cliffs, preferably ones in the process of tumbling down (so that you could dig among loose rocks for a better peek). Winter was the best season for fossils, despite the cold and wet, because storms swept in and battered the cliffs and, sometimes, exposed new rocks.

Summertime visitors knew little of those hazards. For them, Lyme Regis was a charming spot with striking views and a bustling social scene built around card playing and dancing. Jane Austen took approving note of the "bathing machines" that dotted the beach, for bashful bathers.

To this day, warning signs on the cliffs at Lyme Regis—scarier versions
of the drawings that show someone slipping on a wet floor in the
supermarket—show how quickly fossil hunters can run into trouble.

Few people at the time knew how to swim. But "bathing machines"
were not meant for swimming, in any case, but for healthful plunges
into the sea. Swimming was looked on with distrust, as less a sport than
an open invitation to drowning. School primers included such sample
sentences as "Many swim in, but swim not out again."

The "machines" were simply wooden huts on wheels that horses
pulled into shallow water. (Men, who did not need to be as coy as women
about showing their bodies, generally bypassed the huts and simply
splashed about where they liked.)

With the hut turned away from the beach, a bell rang out to warn
swimmers and boaters to keep their distance. The bather opened the
door and climbed down a few steps into the water, where professional

dippers stood at the ready. These were robust women whose job was
to help bathers dunk under the waves and rise back up, sputtering but
invigorated.

Then it was back into the bathing machine for the short ride to the
beach and a chance to change back into presentable clothes. (Another
century would pass before women ventured onto the beach in their
bathing outfits.)

This cartoon, from 1813, shows men ogling women as they spilled
from bathing machines into the water. Reality was far tamer
than in Thomas Rowlandson's imagination; women wore modest,
cumbersome bathing costumes.

For the locals, life was harsher. "The country people here are wretch-
edly poor," wrote William Wordsworth in 1797, in a house he had rented
a few miles from Lyme. Three years later, the harvest failed and famine
struck. Riots broke out, and the government called in troops to restore
order. Richard Anning was among the leaders of those clamoring for
bread.

Scholars who revisited this era many decades later invoked bland
terms like *modernization*, but the changes were cruel for those caught

in history's gears. Farmers were thrown off their land and ended up as hired hands on large estates. Small mills that had once produced cloth and lace in southern towns like Lyme were made obsolete by huge factories in northern, industrial cities like Manchester. Prices fell, but, at least in the south, jobs vanished.

The Annings did their best to scrape by, with fossil sales here and there supplementing Richard's meager carpentry income. Then, in 1810, the picture grew even darker. Richard died unexpectedly at the age of forty-four, the victim, in one historian's unsentimental summary, "of the combined effects of having fallen over a cliff on his way to Charmouth, and of consumption."

Mary Anning, eleven years old, was fatherless, and the family was deep in debt.

CHAPTER 3

"The Most Amazing Creature"

The day after her father's death, as Mary later told the story, she had been wandering along the seashore when she found an ammonite. A lady spotted her carrying her prize and offered to buy it for two shillings and sixpence (about two days' wages for a workman). From that moment on, Anning recalled, she was "fully determined to go down 'upon beach' again."

It was, at best, a hard living. For the next five years the family would be on relief. Life was grim, even with that help, and it remained grim. "I found [the Annings] in considerable difficulty," one visitor reported in 1819, "in the act of selling their furniture to pay their rent, in consequence of their not having found one good fossil for near a twelvemonth."

But during those lean years, Mary Anning had made a stupendous find. This would be her first great coup, but her older brother deserves a great deal of credit, too.

On a summer day in 1811, Joseph Anning spotted something odd on the beach under a heap of dark rocks. (Joseph was fourteen, Mary twelve.) He dug free what looked to be a giant skull, perhaps four feet long, with hundreds of sharp teeth and two enormous eye sockets the size of dinner plates. *A crocodile? A giant fish?*

The skull was brownish black and looked uncannily like a statue that had been carved in driftwood long ago and then abandoned for time and weather to do their work. (Living bones are white, but fossilized bones are dark. The coast near Lyme Regis still draws novice fossil hunters, along with scads of experts, and the novices tend to imagine they will see something akin to a white, bony rib poking out from a cliffside. A real find would likely be far better hidden, and a newcomer would almost surely miss it. Perhaps a dark stone might feature a small, out-of-place bump or a tiny crack that offered a glimpse of a concealed bone inside.)

The skull's size was startling, and so were its teeth. Even today, when the skull is locked away in a museum's display case, it looks fearsome.

It is hard to watch the paddleboarders and swimmers happily enjoying the beach at Lyme Regis today without imagining this giant beast patrolling the seas, cruising beneath the waves, awaiting its chance. Perhaps it is the oversized eye sockets that are most unnerving. "An eye," one nineteenth-century scientist said with a shiver, "sometimes larger than a man's head."

Joseph convinced two local men to help him carry the skull home. He showed Mary where he had found it, but they searched in vain for more of the beast's bones. It took Mary more than a year before she found another bit of buried bone, not far from where the skull had lain but about two feet underground.

Near pay dirt at last, she dug on. A glimpse of a black, knobby bit. Backbone. Ribs. More backbone. The words *fossil hunter* make it sound as if bones might be lying in plain sight, like discarded bottles by the roadside. But fossils are typically embedded in solid rock with perhaps a small bit poking out, and spotting them and then coaxing them free can be the labor of years. Mary chipped away with her hammer and chisel.

In time she recruited several men to help her. Their prize, finally, was an enormous dolphinlike creature, seventeen feet long and with its skeleton still intact. With huge jaws and daunting teeth, it had plainly been a formidable predator. It was put on display in London, where scientists gawped—one viewer proclaimed it "the tyrant of the deep"—but had no idea what to think.

Scientists dubbed the find ichthyosaurus ("fish lizard"). Mary Anning had made the discovery, but she had no say in its naming. Names were bestowed by scientists, not unlettered girls. But a local landowner paid £23 for her find, and that was money enough to put food on the table for six months.

The skeleton eventually ended up in the Natural History Museum in London. More accurately, the skull is on display. Somehow the rest of the body, which Mary dug up with so much trouble, has been misplaced. But the museum does have on display a different and nearly complete ichthyosaur skeleton, "one of the largest and most complete known," which Mary found two decades later.

In the meantime, scientists poring over fossil collections made a tantalizing find. Some of the bones looked as if they belonged to ichthyosaurs, but not quite. Had *another* marine reptile shared ancient seas with ichthyosaurs?

In the winter of 1820–21 Mary Anning took a giant step toward solving the mystery. She found a skeleton of a new marine creature, and it was definitely not an ichthyosaur. The strange new fossil was missing its skull but otherwise was nearly complete. Anning and others kept hunting. Finally, in 1823, success!

Anning had unearthed a complete skeleton of a creature that one scientist dubbed "altogether the most monstrous" ever found. This was a huge sea-dwelling reptile now called a plesiosaur. At age twenty-four, the untutored young woman had *two* great discoveries to her credit.

(In everyday speech, all huge prehistoric creatures are "dinosaurs," but the word has a technical definition having to do with particular details of anatomy involving the hips and backbone. Mary Anning's plesiosaurs and ichthyosaurs lived two hundred million years ago, but they were not dinosaurs. It was not until 1858, after Anning's day, that fossil hunters found dinosaurs in Lyme Regis.)

The ichthyosaur looked more or less like a huge, ferocious fish. No one could quite think what a plesiosaur looked like. Nine feet long and four feet across (from paddle tip to paddle tip), with a ludicrously long neck and tiny head, the bizarre-looking creature left onlookers stammering in confusion. One expert suggested that it bore some resemblance to "a serpent threaded through a turtle."

The geologist William Buckland opted for an even odder amalgam. "To the head of a Lizard," he wrote with a mix of fascination and horror, "it united the teeth of a Crocodile; a neck of enormous length, resembling the body of a Serpent; a trunk and tail having the proportions of an ordinary quadruped, the ribs of a Chameleon, and the paddles of a Whale."

This meticulous drawing, by Mary Anning, shows her plesiosaur. Collectors contacted her at once, hoping to purchase her find. "One thing I may venture to assure you," Anning wrote in this letter to a would-be buyer. "It is the first and only one discovered in Europe."

So unlikely was the creature that scientists believed at first that it was a hoax. The eminent Georges Cuvier, the French paleontologist, suggested that the skeleton might have been assembled from different creatures. Had Anning been taken in by a forgery? Or perpetrated one?

Cuvier came around when William Buckland and William Cony-beare, both of them distinguished scientists, vouched for Anning. They sent Cuvier detailed drawings and reminded him that it was Anning who had unearthed an ichthyosaur skeleton years before.

Cuvier recanted. He was an imposing figure, brilliant and accomplished and weighed down with honors. Few had the intellectual clout or the political savvy to stand up to him. ("A man of devouring ambition, a master diplomat, one of the handsomest men of his day, and iron-willed," in one biographer's judgment, "Cuvier was ruthless in ridding himself of opposition.")

Georges Cuvier

Now this one-man Supreme Court delivered his verdict on Mary Anning's plesiosaur. Her unlikely skeleton—"the most monstrous assemblage of characteristics that has been met with among the races of the ancient world," in Cuvier's judgment—was not a blunder or a scam. Far from it. Cuvier proclaimed the find a triumph: "It is the most amazing creature that was ever discovered."

On the evening of February 20, 1824, at the annual meeting of the Geological Society, the world was introduced to the plesiosaur. William Conybeare, a reverend and a geologist, delivered the news. "We adjourned to the Society's rooms at 1/2 past eight," he wrote to a geological colleague, "and there I lectured on my monster."

The plesiosaur was "his," in Conybeare's mind, because he had been one of the first to propose that ichthyosaurs had shared the sea with a second strange, fishlike beast. Before Anning made her find, he'd examined several odd-looking vertebrae that supposedly came from an ichthyosaur. He didn't buy it. Then Anning turned up, and her evidence confirmed his suspicions.

Conybeare was not a good speaker, but his talk was a smash. He relied on a careful drawing of the skeleton that Anning had provided. (The plan had been to show the skeleton itself, but that hadn't worked out. Anning had shipped her prize skeleton to the Geological Society in an enormous crate, but ten men had spent a day in futile efforts to carry it upstairs to the society's meeting rooms.)

Conybeare wowed his listeners with descriptions of these unlikely, long-necked creatures. "They were as alien to his audience as if they had hailed from another planet," writes the paleontologist Christopher McGowan, and the spectators sat entranced.

Anning was not among them; the Geological Society would be closed to women for almost another century, until 1919. At the time of the talk, only two plesiosaurs had ever been found. Mary Anning had found both of them. Conybeare never mentioned her name.

CHAPTER 4

An Epic Written in Chalk

Mary Anning's finds were daunting in two ways at once. The bones and skeletons that she and other fossil finders unearthed were eerie in their own right. Even worse, the unlikely settings of those finds delivered a shock of their own.

When bones and teeth and similar relics turned up, they were nearly always deep underground, buried beneath thick layers of rock. They might well have stayed lost forever if not for a fluke—a quarry cut deep into the shale, perhaps, or a storm that eroded away a cliffside.

How had creatures come to be entombed inside solid stone? They surely couldn't have *tunneled* in, as prairie dogs dig through the soil to fashion their burrows. Perhaps the poor beasts had been caught in a lava flow or trapped in tar and buried alive?

Elsewhere, perhaps, but geologists could tell that *these* rocks had formed ever so slowly, in layers, over the course of eons. So the rocks were ancient, which meant that the relics trapped inside them were ancient, too.

They had not been captured in an instant, like mice in a mousetrap. Instead, they had been entombed in slow motion, after their deaths, as if ever-so-tiny gravediggers had been at work for millennia, heaping grains of sand and teaspoons of clay and mud atop a corpse.

It fell to the geologists to make sense of these discoveries. This was not a mission they had intended. Their goal was to learn about the structure of the earth, not to find relics of ancient life. They set out to study rocks and, in effect, tripped over cadavers.

When the telescope was invented, around 1600, humans felt a kind of intellectual vertigo as they tried to grasp the news that Earth was not at the heart of the cosmos but merely a dot in a nondescript suburb. When geologists delivered word that time was a vast expanse, too, vertigo took hold once more.

The geologists offered up their news around 1800. Most people at the time still believed that the world was six thousand years old, as the Bible taught. One of the famous poems of the age, for instance, described the ancient city of Petra, in Jordan. Carved from sandstone cliffs, Petra was "a rose-red city half as old as time."

To modern ears the line sounds appealingly romantic but impossibly vague, like "from here to eternity." To the poet John Burgon and his readers in Victorian England, "half as old as time" was as straightforward as "halfway from London to Paris."

Earth had been created in 4004 BC (on the evening of October 22), according to the most widely cited figure. Halfway there put Petra at about 1080 BC, which fit nicely with beliefs at the time.

Nowadays we picture six thousand years as a blink of an eye, geologically speaking. We think of sharp, craggy mountain peaks eroding to rolling hills, or rivers carving canyons, and it seems certain that such changes could not be squeezed into a few thousand years. But that thought is new.

For ordinary people in 1800, *six thousand years* conjured up thoughts not of confinement and constraint but of vastness and eternity. Six thousand years corresponds to about two hundred generations. Human memories last perhaps three generations—who can imagine their grandmother's childhood? On a human scale, two hundred generations is effectively "forever."

Then came the geologists with the astounding news that Earth's age was measured not in thousands of years, but in tens of thousands or millions or perhaps many millions. That was a revelation, but it had an utterly mundane origin, as far as could be from a lightning bolt crackling across the sky. It was sedimentary rocks—rocks formed from layers of sand or mud compressed into a solid mass—that had opened the geologists' eyes.

Take the White Cliffs of Dover. By the 1800s, scientists had recognized that the cliffs were formed from the shells of microscopic sea creatures that died and sank to the ocean bottom. The creatures, called coccoliths, are absurdly small. A heap of several hundred would occupy the volume of a single grain of sand. But the cliffs stand hundreds of feet high, on England's southeast coast. They form what is, in effect, the world's biggest piece of chalk. On a clear day you can see the White Cliffs from France.

Only time—eons heaped upon eons—could have made that happen. As the minuscule shells of those microorganisms drifted down from near the ocean's surface to its floor—day after day, decade after decade, century after century, millennium after millennium—they formed a graveyard of infinitesimal broken bits. This corner of England is, in the words of one paleontologist, a "countryside made of skeletons."

Each newly arrived shell added a tiny downward push to the pressure on the shells already in place. (A deep enough heap of *anything*, no matter how light—feathers, rose petals, butterfly wings—would eventually grow crushingly heavy.*) More important than that downward, grinding pressure were chemical interactions between the shells and seawater, which glued the bits together.

---

* The Roman emperor Heliogabalus, who was depraved even by Rome's standards, supposedly used a cascade of rose petals as a murder weapon. To amuse himself, Heliogabalus invited guests to a banquet and then released a torrent of rose petals from a false ceiling to smother them. But their deaths were presumably due to suffocation, not to the weight of the petals.

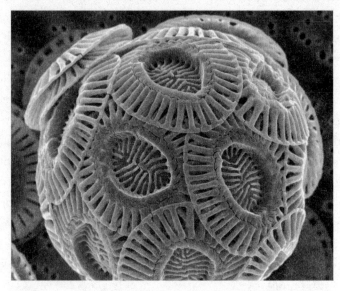

Chalk, as seen through a microscope. The discs are infinitesimally tiny scales that come from the bodies of sea creatures called coccoliths.

Modern-day scientists have filled in the rest of the story. Twenty or thirty million years ago, Africa rammed into Europe. The Alps rose into the sky, and, further west, vast chalky deposits that had been at the bottom of the sea were heaved upward. Then came a long period of peace and, finally, another cataclysm. This was mere hundreds of thousands of years ago, when a vast lake near what is now the North Sea burst through a natural dam that had held it back.

A mega-flood with a surge equivalent to one hundred Mississippi Rivers roared its way south. At the time, an immense land bridge connected England and France. The racing waters sliced through that bridge and carved the English Channel, turning England into an island and revealing the White Cliffs.

The cliffs stand silent now, but looked at with a geologist's eye, they tell a story as dramatic as the *Odyssey*. In the mid-1800s Thomas Huxley delivered a famous lecture on the message of those white cliffs. "A great chapter of the history of the world is written in the chalk," Huxley observed.

*   *   *

The story of one discovery shows just how difficult it was for non-geologists to come to terms with the new ideas about the vastness of time, even as late as the middle of the nineteenth century.

The first Neanderthal bones turned up in a cave in Germany in 1856. Those few bones—a chunk of skull and some leg bones—were immediately recognized as human, although decidedly strange looking even so. The oddest feature was a thick, prominent bony ridge in the skull, just above the eye sockets.

We now know that the bones belonged to a man who had been dead for forty thousand years. But two of the most eminent scientists of the day argued strenuously that he had been dead only *forty* years.

Today the debate could be resolved at once. But the modern techniques rely on discoveries that had yet to be made—involving a wholly unsuspected phenomenon called radioactivity, above all—that would have seemed, in the mid-1800s, as far-fetched as invisibility cloaks.

Instead, Rudolf Virchow and August Mayer assigned an age to these curious bones by careful detective work. Virchow chimed in first. He was brilliant, opinionated, and a leading figure in every scientific and political controversy of his day (he explored Troy with Heinrich Schliemann; he wrote two thousand scientific papers; he fought in the streets of Berlin in the Revolution of 1848; and he was the first person to realize that cancer arose when a healthy cell turned rogue). Virchow was a heavyweight.

When he was called on to examine the Neanderthal bones, he pointed out the unusual curvature of the thigh and pelvic bones. To the trained eye, Virchow explained, this was plainly the skeleton of a modern human being who had suffered from rickets, a bone disease.

August Mayer, a much-admired anatomist at the University of Bonn, added more detail. This was indeed the skeleton of a modern human, he agreed, and its unusual appearance offered clues to the man's identity. The shape of the bones testified to a lifetime spent on horseback. Mayer pointed out, also, that the skeleton's right arm was broken.

Afflicted by both disease and accident, then, this poor man had endured excruciating pain. That was the key. Why did the skull have an overdeveloped, bony ridge above the eye sockets? Because unceasing pain and a lifetime of wincing would have made for a perpetual furrowing of the brow.

Fitting the puzzle pieces together, the two scientists named their suspect. The skeleton likely belonged to a Russian Cossack who had died battling Napoleon's troops, in 1814.

Even at the time, this story might itself have caused some furrowing of the brow. How had the tormented cavalryman dragged himself deep into a cave in Germany, for starters? And how, in a matter of a few decades, had he come to be buried under six feet of sedimentary debris?

But awkward questions like those didn't turn out to be deal-breakers. (We now know, incidentally, that Neanderthals happened to have thick, bowed leg bones.) As far-fetched as it was, the Cossack theory had one great advantage in its favor. It made the world a bit more familiar, a bit more understandable.

Confronted with mysterious relics from a strange-looking human, you might have wondered if the human story had begun unimaginably long ago. You might have wondered if humankind's family tree had ancient roots and unfamiliar faces.

But that would be asking a great deal. More likely, you would have opted for a more soothing explanation. Perhaps a soldier had ridden out on horseback a generation or two before and lost his way.

Small wonder, then, that when scientists first stumbled upon *dinosaur* bones—from unknown creatures that lived a hundred million years ago—they stared in slack-jawed confusion.

CHAPTER 5

"The Dreadful Clink of Hammers"

B y now we have had two centuries to learn to navigate the abyss of time, and still we have scarcely managed it. Scientists today figure that Earth is 4.5 billion years old, rather than six thousand.

The revised figure itself is only part of the point. Just as important is the vastness of the change. It is not merely that history expanded from short to long; the point is that it expanded to an almost unfathomable degree.

We blithely toss around references to *millions* and *billions*, but the familiarity of the words obscures their meaning. Take the change from six thousand years to 4.5 billion. If a tennis court grew by the same proportion, the new court would stretch from the North Pole to the South Pole and the net would be at the equator.

Imagine keeping your bearings in a world that had changed to that extent.

The expansion of time was only one aspect of the problem. Even worse, the new notion of time forced a new understanding of our familiar home and our place in it. For two thousand years, scientists and theologians had agreed that humankind occupied a special, distinguished niche in creation. "Man stood at the center of all things and the entire universe had been created for his edification and instruction," one historian writes. "Hills had been placed for his pleasure, animals ran on four feet because it made them better beasts of burden, and flowers grew for his enjoyment."

At least in the Western world, the belief that nature existed to serve human needs was almost universal. (Primers drummed the message home from a pupil's earliest days in school. The tone of *Lessons on Natural Philosophy, for Children*, from 1846, was typical. "Can you tell what good the wind does?" the author asked. "It blows off bad air from places where there is dreadful sickness," and "it blows ships across the sea.")

How could it be, then, that entire eras had come and gone before humans had ever appeared on the scene? And that throughout those long ages, creatures had devoured one another or cowered in the shadows, and all those dramas had played out unwatched by human eyes? None of that made any sense. Who would put on a play in an empty theater?

Even putting dinosaur bones to one side and thinking only of less dramatic fossils, the same perplexing riddle swam into view. "Why so much beauty when there was no eye of man to see and admire?" one of the best known of the early geologists wondered, in a letter to a friend in 1834. (He had in mind creatures akin to the chambered nautilus.)

"Does it not seem strange that the bays of our coasts should have been speckled by fleets of beautiful little animals, with their tiny sails spread to the wind and their pearly colors glancing to the sun, when there was no intelligent eye to look abroad and delight in their loveliness? Of all the sciences there is none which furnishes so many paradoxical facts and appearances as geology."

These were tantalizing riddles, and everywhere you turned, you confronted dozens more. New discoveries about the age of Earth and its unlikely inhabitants shredded old certainties. John Ruskin, the great art critic of the Victorian age, wrote that his faith "fluttered in weak rags."

Science was the culprit. "If only the Geologists would leave me alone, I could do very well," Ruskin moaned, "but those dreadful Hammers! I hear the clink of them at the end of every cadence of the Bible verses."

Ruskin spoke for countless others struggling to find their way in a landscape where the signposts had all vanished. Suddenly scientists and laymen learned that time extended back almost forever, and that strange creatures had once ruled the world and then vanished, and that the world of the past looked nothing like the one we know.

If nature was benign, as Paley taught, what was the message of those newfound, giant bones? Plainly Earth had once been home to immense creatures. Where had they gone? Could *we*—we who were, after all, the point of creation—suddenly vanish, too, popping out of existence like soap bubbles? *Wait—what?!*

Questions about life and nature and purpose all came under the broad heading of "evolution." The notion was in the air in the early decades of the 1800s, although Charles Darwin had yet to step onstage (*On the Origin of Species* wasn't published until 1859). This was not evolution as we now think of it. There was no talk of "survival of the fittest" or "the struggle for survival."

In these early days, *evolution* meant "unfolding" or "unrolling." This was a nonthreatening, even an encouraging, view. In the cheery summary of the nineteenth century's most famous preacher, Henry Ward Beecher, "the whole physical creation is organizing itself for a sublime march toward perfectness."

Not everyone agreed that plants and animals evolved—many still held to a static, diorama picture of creation—but even those who embraced change pictured life as advancing slowly and steadily and always on an upward trajectory. (In ordinary speech we still use *evolved* in this sense. "My father was the worst—you should have heard him—but he's more evolved now.") Instead of chance and happenstance, the themes were design and order.

From simple beginnings, these nineteenth-century pioneers believed, nature had advanced to ever more sophisticated designs, much as carriages had progressed from wheelbarrows to wooden carts to elegant horse-drawn coaches. In the case of the natural world, the path led from snails and worms at the base of the pyramid of life to humankind at the peak.

The early 1800s saw no need for a Darwin to explain the mystery of the world, because no one saw a mystery. The question *Why is the natural world so perfectly arranged?* was a good one, but the answer was obvious. The world was an elegant and harmonious machine because the Creator built it that way.

But Darwin will not appear in our story until the end, and our scientists have no idea that for twenty years a reclusive genius has been gathering explosives to blow their world apart.

Questions about design and meaning would have been fraught in any era. But when dinosaurs barged their way into the debate, even abstract ideas took on vivid, snarling form. The geologists' declaration that Earth was old, not young, was worrisome. But it was too remote to pack an emotional punch, just as astronomers' declarations today that there are a hundred billion stars in a galaxy fail to bowl us over.

We *would* perk up if someone showed up with incontrovertible proof that aliens had traveled from one of those stars and paid us a visit. A video of an alien army or a bit of alien corpse that matched nothing on Earth would be hard to dismiss. For our forebears in the nineteenth century, bones and skeletons from fierce, extinct creatures served as that sort of impossible-to-explain-away evidence.

All educated Victorians were steeped in the classics, so they knew the story of the triumphal parades held to honor a commander who had come home to Rome victorious in battle. As the cheers of the crowd rose up, a companion whispered in the great man's ear, "Remember you are mortal."

In the nineteenth century, geology and paleontology took on that whisperer's dread role. With its fossils and its extinct creatures and its vanished worlds, science delivered the fearsome message "You, too, are mortal. You, too, will pass away."

CHAPTER 6

"It's a Beautiful Day and the Beaches Are Open"

Prehistoric creatures like Mary Anning's had star power from the start, but the allure was complicated. These were beasts that were simultaneously ancient and brand-new, and no one quite knew what to make of them. In early accounts, dinosaurs and their ilk were "monsters" but then, in the next breath, "wonders" and "marvels."

Most often, scientists played up the horror and soft-pedaled the wonder. Scientific tracts could sound as overwrought as Gothic novels. Fossilized bones were not mere relics of the past, akin to coins or vases, but keys to a bygone world that was home to "myriads of 'creeping things.'"

Collectors gaped at "the fleshless bones of the primitive race of preying monsters" and shivered as they gazed upon "the wide jaws that first committed murder."

William Buckland, one of the most important scientists of the early 1800s, veered back and forth in his portrayal of ancient creatures. On the one hand, the elegance of their design testified to God's careful, loving handiwork. On the other hand, when Buckland described the first pterosaur found in England, his thoughts turned not to a caring, meticulous Creator but to Satan himself.

Buckland described the prehistoric reptile's wings and claws and pictured it flying across the sky, and scuttling over the ground, and climbing into trees. He shuddered, and then he quoted Milton's description of the devil in *Paradise Lost*: "The Fiend . . . With head, hands, wings, or feet, pursues his way, / And swims, or sinks, or wades, or creeps, or flies."

To this day, we have not sorted out how to think of dinosaurs. Time drains many discoveries and inventions of their wonder. Who today marvels at a lightbulb? But as the success of the *Jurassic Park* franchise demonstrates, dinosaurs retain their power to frighten and to fascinate.

For religious believers—and the nineteenth century was a devout age— everything to do with dinosaurs and the depths of time spurred anxiety and confusion. It seemed nearly impossible to reconcile these new notions with biblical teachings. How to make sense of eons of time when Genesis fit all of creation into a mere six days?

What about the mayhem of dinosaurs in bloody combat, when pain and death didn't enter the world until Eve plucked the apple? The dinosaur discoveries, in one modern historian's summary, "shrieked against a straight reading of Genesis."

But publishers figured out early on that readers gulped down accounts of prehistoric creatures. (The devout may have shuddered and clapped their hands over their eyes, but they peeked through their fingers and continued reading.)

Scientific writers sometimes tempered their language—in much the same way, novelists of the era used euphemisms like *fallen angel* for charged words like *prostitute*—but this made little difference. Even innocent-sounding words like *ancient* were links in a lurid chain, because *ancient* meant "primitive," and *primitive* meant "violent."

Scientific books and articles came with illustrations, which revved up the excitement, all the more so because no one had ever witnessed the scenes the artists depicted. Imagination ran free. Beasts with flailing jaws fought rivals with dagger-sharp claws.

One early encyclopedia of natural history, from 1834, featured some of the first color scenes of extinct animals swooping through the air and prowling the sea. (The colors were invented out of thin air, whereas the shapes of the animals' bodies were at least guesses based on actual bones.)

These bygone creatures were so "strange" and gigantic and fierce, the encyclopedia reported, that they "seemed not so much reality but rather the work of a diseased imagination."

That tone, with its mix of fascination and uneasiness, was characteristic of the age. The most sophisticated writers were as enthralled as the most down-market hacks. Percy Bysshe Shelley, who had long been captivated by science, conjured up eerie prehistoric scenes featuring bizarre creatures—"unknown winged things" and weird fish, and serpents like "bony chains"—that "once were monarch beasts."

Those uncouth monsters had reigned "on the slimy shores / and weed-overgrown continents of earth," Shelley wrote, and in that hideous setting had "increased and multiplied like summer worms / on an abandoned corpse."

Dinosaurs were the most vivid, the most *real*, of all such creatures from the past. They truly were, in nineteenth-century eyes, emissaries from another world—they were aliens made flesh (or, actually, bone). But as time passed, outright fear gave way to edgy curiosity.

Unlike true aliens, who would have been impossibly frightening, dinosaurs were tame ambassadors who could be gawked at in safety. In modern times we have defanged the notion of "aliens" by turning them into "little green men." In the nineteenth century, "big green animals" worked the same trick.

For the first generations to confront the reality of dinosaurs, much of the fascination with the towering creatures had the same tangled roots that it does for six-year-olds today. These were the ideal sort of monsters—big, scary, and, best of all, dead.

But dinosaurs remained unsettling, even so, simply because they were so utterly nonhuman. We cringe when we cannot connect. (Sharks terrify us not just because of their huge, razor-sharp teeth but also because of their dead, black-button eyes.) Dinosaurs are indeed safely distant from us, but they are distant, too, in the sense that we can never forge a bond. And every day brought word of new bones, new teeth, new skeletons.

Nineteenth-century scientists found themselves in the predicament of the mayor of Amity in *Jaws*. As the decades passed and the evidence piled up—as it grew ever harder to deny that time stretched back into an endless past, and that countless species had vanished from Earth, and that humans were latecomers to the story—they dashed around frantically trying to reassure everyone (and themselves) that all was well and the world was orderly.

But the sound of the cellos grew ever louder.

CHAPTER 7

Trembling in the Dark

Since no one has ever seen a dinosaur, no one can be certain what they looked like. Even today the best guesses can vary widely, like police sketches in the early days of a manhunt. The most famous dinosaur of them all, Tyrannosaurus rex, emerged from a recent makeover at the American Museum of Natural History with bright orange skin and a mane of feathers.

Showy redos like that are a product of the past few decades. Dinosaurs have recently acquired feather boas and feathered crowns and stripes in shimmering reds, whites, browns, and blacks. The glitz makes an impact, like designer gowns on the red carpet, but no one knows exactly what role it played.

Dinosaurs could likely see a great many more colors than we can, because reptiles and birds can. (Humans have three kinds of cone cells in our retinas; reptiles and birds have four. By the same token, we can see vastly more colors than dogs can, because they have only two kinds of cones.) The dinosaurs' world would have shimmered with colors invisible to us, and quite possibly so would their skin.

The specific ideas about the colors of dinosaur feathers come by way of birds. In birds' feathers, tiny packets called melanosomes serve as pigment-containing pouches. The packets come in different shapes, each shape corresponding to feathers of a particular color.

The story about bird feathers and melanosomes is old news, but 2008 brought a thrilling breakthrough—scientists found melanosomes in fossilized *dinosaur* feathers, which presumably meant that dinosaur feathers came in colors, too. Even better, the melanosomes in dinosaurs came in the same telltale variety of shapes that they do in birds, so you could guess exactly *which* colors dinosaurs flaunted.

Dinosaurs would stand out in any lineup, though, no matter their colors. They truly shook the earth; the biggest weighed as much as ten elephants. One colossal dinosaur, discovered in China in 1952, had a fifty-foot-long neck, a neck longer than a city bus.

Other prehistoric creatures were huge, too. Ancient dragonflies had the wingspan of hawks. Even millipedes—the ancestors of the little roly-poly bugs that turn up in leaf piles today—could grow to nine feet long and one hundred pounds. Scientists have yet to agree on why life today is on so much smaller a scale. Once-popular explanations—the atmosphere was much higher in oxygen in prehistoric times than it is today, for example—seem not to have held up.

Turn from dinosaurs' size to a different measure, and you find a different superlative. The great white shark is the animal in today's world with the strongest jaws. Tyrannosaurus rex's jaws were twice as strong. "Could T. rex have bitten a car in half?" one modern paleontologist asks. "The answer is a resounding 'yes.'"

And dinosaurs had an unfathomably long run; they reigned for over one hundred million years. (Some scientists believe the true figure is closer to one hundred eighty-six million years.) Modern humans have been around for perhaps a hundred thousand years. If humans manage to survive ten times as long as we have so far, we will have made it 1 percent as long as the dinosaurs did.

That reign was so long that it's worth looking at from more than one angle. Here's another way to try to make sense of a span of one hundred million years. We think of dinosaurs as if they all lived at the same time, but that's not correct. The ones who lived early on had gone extinct by the time the latecomers showed up.

Take stegosaurus. With a row of upright plates along its back, it is one of the most familiar dinosaurs of all, nearly as recognizable as T. rex. Every set of toy dinosaurs includes a stegosaurus.

Reference books all tell us that stegosaurus had come and gone before the first appearance of T. rex. But that bare fact skips past the real surprise: the span of time from stegosaurus to T. rex was longer than from T. rex to the iPhone.

The dinosaurs were finally done in by an asteroid that crashed to earth sixty-six million years ago. It knocked over life's gameboard, and not just for dinosaurs—75 percent of all the species of plants and animals on Earth were killed, and so were 99.9999 percent of all the individual organisms.

The asteroid was the size of Manhattan and traveled twenty-five times as fast as a bullet. The crash triggered earthquakes and tsunamis. Mile-high waves raced across the ocean at the speed of an airliner. Countless tons of pulverized, molten rock shot into the air. Some of the rocks were flung to the moon.

As debris rained down—and much of it was hotter than the surface of the sun—it ignited forest fires around the globe. Air temperatures reached five hundred degrees, as if the entire planet were in an oven set on "high."

Soot and dust enveloped Earth. With sunlight blocked for months on end, all but a fortunate few plants and animals died. Ferns and mosses survived. So did sharks and turtles, crocodiles, frogs, and dragonflies, and a handful of small, scurrying mammals.

It was a cosmic fluke that did in the dinosaurs, not a dead-end design. We seldom picture things that way. It seems more natural to think of the story of life as a sort of Olympics, with humans perched on the victory stand with a gold medal around our necks. Here is a case, surely, of merit receiving its proper reward.

But if we can imagine a dinosaur taking a pen in a mighty claw, it would no doubt write the story in a different way. For vast eternities, dinosaurs ruled Earth, and mammals scuttled about in the dark, like rats, hoping not to get squashed underfoot. (Even with the dinosaurs gone, two hundred thousand years would pass before mammals first dared to venture out in daylight.)

If not for a cosmic accident, our rodent-sized forebears would presumably still be there trembling in the moon's faint light, and humans would never have come to be.

And so our rise seems less a matter of a hard climb to the top and more akin to the story of a mailroom clerk who suddenly finds himself CEO because nearly everyone else was killed in an earthquake.

Chance played a key role in the dinosaur story. We know that now. But nineteenth-century scientists had no such belief. They rejected chance because they took for granted that God had matters well in hand. That focus on God might seem surprising—what did God have to do with science?—but that only highlights how different the world of our forebears was from our own.

God figured in every scientific debate, because evidence of design and planning was all around. The only alternative that anyone could imagine seemed not only far-fetched but also grotesque. In the fourth century BC, the Greek philosopher Empedocles had suggested that the first humans had arisen by chance. Before that, random body parts had somehow wandered about on their own.

"Many heads grew up without necks," Empedocles wrote, "and arms wandered about naked, bereft of shoulders, and eyes roamed about alone with no foreheads." Those parts combined at random, yielding animals with human heads, or humans with animal heads, or creatures with two faces.

But occasionally, by good fortune, the pieces came together in an appealing, efficient way. The mismatches soon died off, Empedocles explained, and the world was left to the fortunate few and their equally well-formed descendants.

Even in ancient Greece, thinkers had hurried to shout down that proposal. As centuries passed—and as Christianity and a belief in an all-powerful, all-knowing God took root—the hostility toward any doctrine built around randomness grew ever more fervent.

By the nineteenth century, the notion of God as designer was entrenched almost too deeply to question. The balance of nature served as a favorite example.

Why does the world look roughly the same from year to year, with some animals common and others rare? Why are there more antelopes than lions, more rabbits than hawks? Because God is a divine planner who has carefully done his sums and arranged things so that they work perfectly.

This seemed certain in the early 1800s, so obvious as scarcely to require demonstration. Today an altogether different answer seems just as plain. If lions are made from the flesh of antelopes, and lions are big and antelopes are small, then—so long as conditions stay about the same—there will be few lions and lots of antelopes.

So any modern ecologist would explain. (One classic ecology text is entitled *Why Big Fierce Animals Are Rare*.) All the more so when you recall that not every bit of antelope can be transformed into lion; it takes work to break down and reassemble that flesh, and inevitably there is waste along the way. We would no more drag God into this story than we would if called on to explain why it takes many bricks to build a single house.

But, as our intellectual ancestors saw it, they had not "dragged" God anywhere. How could they, when his natural place was at the heart of every story?

CHAPTER 8

The Divine Calligrapher

The dinosaur discoveries came out of nowhere, like the asteroid, and the public in the nineteenth century was scarcely better prepared than the dinosaurs had been. It was not just that such things as monstrous skeletons were contrary to experience. The shock was that they were contrary to reason. Such things *could not be*, because they had no place in a world that was, everyone knew, under divine supervision. Why would God have indulged in such follies?

In many ways the nineteenth century had begun to look like the world we know, with railroads and sprawling cities and factories pouring smoke into the air. But a vast gulf separates what we believe, in this secular age, from what our forebears took for granted. "The most important thing to remember about religion in Victorian England," one historian notes, "is that there was an awful lot of it."

It seeped its way everywhere. In the 1840s and '50s, for instance, scientists took up the question of whether there was life on other planets. *Impossible!*, thundered William Whewell. What sense did it make to assert that other worlds might teem with life—had Jesus died for *their* sins?

This was not the view of an eccentric or a fringe thinker. Whewell (pronounced HUE-ul) was an eminent scientist with impeccable establishment credentials (and it was Whewell who coined the word *scientist*).

Equally illustrious thinkers opposed Whewell. But they, too, made their case on religious grounds. If there was no life elsewhere in the universe, the Scottish physicist David Brewster declared, then there would be only "lamps lighting nothing—fires heating nothing—waters quenching nothing—breezes fanning nothing—everything around, mountain and valley, hill and dale, earth and ocean, all *meaning nothing*."

Brewster's plea was more a wail of existential despair than a scientific argument. Today we take for granted that some things "just happen." Other planets might harbor life; they might not. But what seems like common sense in our day would have been almost inconceivable in the God-soaked 1800s.

That difference in outlook represents a seismic shift. In ordinary speech we use the word *unthinkable* to mean "too horrible to contemplate." But certain claims can almost be unthinkable in the literal sense, because they are beyond the bounds of imagination. That was the status of the idea that things might "just happen" in the early 1800s.

Science and religion had been at odds over the centuries, most famously when the Inquisition put Galileo under house arrest for endorsing a sun-centered model of the solar system. But in England in the early 1800s, the two approaches to the world seemed to fit together tidily.

Nearly every naturalist in Europe in this era was a practicing Christian—many of the leading English scientists were actually clergymen—and they believed with all their heart that in pursuing science they were exalting their Creator. In other eras, science and religion were in conflict. Not here.

The aim of England's scientists, in particular, was to show that religion and science were two sides of one coin. They clung to that idea as they labored away, for decades. But the next generation of scientists took a different line altogether. Religious faith was all very well as a personal matter, they maintained, but when it came to science, religion only got in the way.

*   *   *

In the early 1800s, both America and Britain still held the happy view that science and religion belonged together. Edward Hitchcock, the first scientist to examine Pliny Moody's dinosaur prints, was both a reverend and a geologist. The two subjects fit together so naturally that Hitchcock's title at Amherst College was "professor of natural theology and geology."

Sir Humphry Davy, perhaps the best-known scientist in the first half of the 1800s, filled his notebooks with poems hailing God's handiwork:

> The eternal laws
> Preserve one glorious wise design;
> Order amidst confusion flows,
> And all the system is divine.

The poetry was labored, but the homage was sincere. Davy's voice was only one in a scientific choir. Religion shaped every debate, and no one doubted that God watches every sparrow that falls.

Geologists and naturalists loudly proclaimed their faith in an all-knowing Creator. To spend one's days studying Earth's mountains and valleys and birds and beasts and then to skimp on praising God would be ungrateful and irresponsible.

To disagree would make no sense, as if someone were to suggest that you might admire a painting and neglect the painter. Could there be anyone so blind, one geologist asked in 1839, as to "survey the beauties of His hand, without a single recognition of Him whose honor and glory they so unceasingly declare"?

This was not to say that all scientists in the nineteenth century read the Bible literally, although some did. But to recognize God's majesty was not the same as to look to the Bible for scientific insights.

By 1800, in fact, few scholars still scrutinized the Bible in search of truths about the workings of the physical world. That was new. A century before, Isaac Newton, who possessed a mind as powerful as any the world has ever seen, had devoted years of his life to such projects as calculating the precise dimensions of King Solomon's Temple and the exact date of creation (he concluded that God had created the world in 3988 BC).

We think of Newton as the greatest of all scientists. But he spent as much time poring over the Bible as he did contemplating the heavens, and he fervently believed that his biblical work was the more important of the two.

Such faith was common at the dawn of the Age of Science. So devout was Robert Boyle, the greatest scientist of the generation before Isaac Newton, that he never uttered the word *God* without bowing his head. "The two great books—of nature and of scripture—have the same author," Boyle declared, which was to say that the scientist in the laboratory and the scholar poring over the Bible were engaged in a single quest.

God had taken delight and care in adorning plants and animals, Boyle wrote elsewhere, and he seemed almost to picture God with a quill pen in hand. The vibrant colors of a hummingbird's feathers or a chameleon's skin were "Flourishes on the Capital Letters of the Alphabet of Nature."

Throughout the 1700s, this was the nearly universal view. God had written that alphabet, and he had written the Bible, and it was impossible to think that the two messages could conflict.

Even as early as Boyle's day, though, many scientists had begun to change tack. Newton, among the most gifted mathematicians of all time, took for granted that God was a mathematician, too. The path to understanding creation went by way of geometry and calculus.

But scientists of a less austere frame of mind than Newton shifted their focus to the living world. The picture of God as a clockmaker meticulously crafting gears and cogwheels gave way to God as a graphic artist bursting with invention and sketching at breakneck speed and with infinite skill. Pages flew off the divine drawing table—here a swooping eagle, there a burrowing mole, a towering elephant, a trembling mouse. God had fashioned wonders at every scale and in every setting—beetles with jewel-like bodies glinting in the sun, fierce cats with bulging muscles and daggerlike teeth, snails and tortoises inching along, cheetahs and hawks moving as fast as the eye could follow.

The shift from God as mathematician to God as artist opened a door. Science grew more inviting. The truths of astronomy were hard to grasp and hard to warm to. But the wonders of the natural world proclaimed God's guiding hand even to those who struggled with the mysteries of arithmetic.

"There is more of admirable Contrivance in a *Man's Muscles*," Robert Boyle wrote in 1688, with the eccentric punctuation of the era, ". . . than in the *Celestial Orbs*; and the Eye of a Fly is . . . a more curious piece of Workmanship than the Body of the Sun."

What was more important than any revision of God's role was what had *not* shifted—the unshakable belief that God had arranged every facet of the world. The historian David Wootton summarizes what virtually everyone took for granted until deep into the 1800s. "A book requires an author; a palace requires an architect; a clock requires a clockmaker; and the universe requires a creator."

A proposition so obvious didn't require an argument; the statement *was* the argument. And when it came to living creatures, like eagles or horses or humans, Wootton points out, Victorian-era scientists saw the case as even stronger.

Palaces were complex, but they were inert. Living, moving, growing organisms were plainly far more sophisticated than even the most ornate arrangements of brick and mortar.

Victorian-era scientists would never have made the comparison, but the way they chose to honor God calls to mind the plaque in St. Paul's Cathedral, in London, praising Christopher Wren, the cathedral's architect: "Reader, if you seek his monument—look around you."

The natural world as scientists in the 1800s pictured it was more orderly than ours, because nothing happened by chance, and it was kinder, too.

This was yet another sign of Paley's influence. His sunny views contrasted utterly with those of earlier religious thinkers.

Look around at the animal kingdom, the Methodist preacher John Wesley had implored, in a sermon in 1781: "What savage fierceness, what unrelenting cruelty are invariably observed in thousands of creatures."

Eve had eaten the apple, and animals as well as humans had been banished from paradise. Wesley worked himself into a hyperventilating frenzy. Was it only the lion and tiger and wolf and shark that "tear the flesh, suck the blood, and crush the bones of their helpless fellow-creatures?"

Far from it. "Nay, the harmless fly, the laborious ant, the painted butterfly, are treated in the same merciless manner, even by the innocent songsters of the grove!"

But it was Paley, not Wesley, whose views took hold in the 1800s. By midcentury, they had seeped so deeply into the culture that a new hymn, "All Things Bright and Beautiful," seemed almost like a précis.

> All things bright and beautiful,
> All creatures great and small,
> All things wise and wonderful,
> The Lord God made them all.
>
> Each little flower that opens,
> Each little bird that sings,
> He made their glowing colors,
> He made their tiny wings.*

---

* Monty Python wrote a spoof, which serves as a reminder of the gulf between the Victorian age and ours. "All things dull and ugly, / All creatures short and squat, / All things rude and nasty, / The Lord God made the lot."

CHAPTER 9

The Apple of God's Eye

The picture of God had changed many times through the centuries. Now, in the early 1800s, came *another* new idea. This one would transform our story.

The new notion transcended the beliefs that scientists and theologians had already come to embrace. It was not just that God had designed every facet of heaven and earth, and that nothing happened by chance, and that the natural world was a gallery displaying his finest work.

Here came an idea more sweeping than all the others—God had made every one of his innumerable choices for the benefit of humans, his prize creation.

That notion of humans as specially favored was new. In the two thousand years that Christianity has been the West's dominant doctrine, its precepts have seldom settled down for long. From the Ten Commandments' "Thou shalt not kill" to the Crusades' "To kill a pagan is to win glory" is only one example.

The shifts follow no clear pattern. Perhaps people eventually tune out a too-familiar message, no matter if it features flames and torture or harps and hymns. When it comes to religion, one historian writes, "Beliefs held almost without question for centuries, and enforced by the authority of venerable institutions, can unpredictably evaporate."

Some early Christians insisted that birthdays should be mourned rather than celebrated, for instance, because all humans are born tainted with sin and the world is a vale of tears. "The saints not only do not celebrate a festival on their birth days," wrote the Greek theologian Origen, around 240 AD, "but, filled with the Holy Spirit, they curse that day."

Birthdays are a small matter, but the question of whether life is a torment or a gift is weighty. Think how far William Paley's vision of a "happy world" was from Jonathan Edwards's message, some sixty years earlier, to his Puritan congregation in New England.

"The God that holds you over the Pit of Hell, much as one holds a Spider, or some loathsome Insect, over the Fire, abhors you, and is dreadfully provoked," Edwards thundered. "His Wrath towards you burns like Fire; he looks upon you as worthy of nothing else, but to be cast into the Fire; . . . he is of purer Eyes than to bear to have you in his Sight; you are ten thousand Times so abominable in his Eyes as the most hateful venomous Serpent is in ours."

Paley, a few generations later, wrote as if he were trying to reassure a listener who had stumbled out of Edwards's church and was still shaking. "The hinges in the wings of an earwig, and the joints of its antennae, are as highly wrought, as if the Creator had nothing else to finish," Paley wrote, his arm comfortingly around his listener's shoulders. ". . . We have no reason to fear, therefore, our being forgotten, or overlooked, or neglected."

Far from forgetting us, Paley's God enjoyed nothing so much as pampering his favorites. Humans need to eat, for instance, and Paley pointed out that an indifferent God would simply have provided us sustenance. Instead he had taken the trouble to make food delicious. "Why add pleasure to the act of eating?" There could be no reason, Paley exulted, "but the pure benevolence of the Creator."

Paley cited dozens of similar examples. Take something as commonplace as a rainy day. What seems ordinary, Paley declared, is in truth miraculous—*water, the very substance we need more than any other, falls unbidden from the sky!* (Paley noted, as well, that it falls not in sheets or rivers, which would be inconvenient, "but in moderate drops, as from a colander.")

Scientists found proof of God's devotion to humankind on every side. William Whewell, highly regarded as both a scientist and a philosopher, looked at the length of the day and saw testimony of God's devotion. Human beings grow sleepy every twenty-four hours, Whewell noted, and it so happens that nighttime rolls around once every twenty-four hours. Could there be clearer proof that God watches out for us?

"If we suppose a wise and benevolent Creator, by whom all the parts of nature were fitted to their uses and to each other," Whewell wrote, "this is what we might expect and can understand. On any other supposition such a fact appears altogether incredible and inconceivable."*

Whewell took his argument to be ironclad because he assumed that the world revolved around humankind. Rather than a mere belief, it seemed a self-evident fact, like *rocks fall* or *water is wet*.

Two centuries later, the world had shifted utterly. For Stephen Hawking, in 1995, the self-evident fact was that "the human race is just a chemical scum on a moderate-sized planet." No nineteenth-century thinkers could have endorsed those words. More important, they could scarcely have grasped their meaning.

To today's way of thinking, Paley and Whewell had the story backward. Humans and other animals look and function the way we do *because* we live in a world where rain is a fact of life. Creatures who were done in by water—squirrels made of Kleenex, perhaps—could never have arisen in the first place.

For us, the nineteenth-century way of thinking calls to mind Aesop's fable about the fly and the chariot. The fly lands on a chariot in midrace and marvels, "Oh, what a mighty dust I have stirred up." (Basketball fans bypass the fable but make the same point on a bumper sticker: *If God isn't a Tarheel fan, why did He make the sky Carolina blue?*)

---

* Another esteemed nineteenth-century scientist noted that God could have set Earth spinning at any rate at all. But he remembered his favorites. If Earth rotated faster than it does, William Conybeare noted, "a gale of wind might, under such circumstances, carry off the roofs of our mansions and every thing which could be torn loose."

We take for granted that Earth would spin or rain would fall just as it does if humans had never appeared on Earth. Nineteenth-century scientists would never have considered so absurd a notion, even to refute it.

And then huge, prehistoric beasts came along and this whole human-centered way of thinking came into question. The problem wasn't accepting the evidence. A jawbone festooned with six-inch-long, dagger-shaped teeth was hard to ignore.

The problem was finding a way to make sense of it. In a world built for humankind and where every feature represented a choice and a decision—where God had painted every stripe on every zebra—where did dinosaurs fit?

What were dinosaurs *for*?

It's worth noting that what struck Victorian England as a mystifying riddle would, in a different time or place, scarcely have upset anyone. Not every culture has believed that humans are the be-all and end-all of creation.

Certainly India and China have not. Even Western societies harbored dissenters. Followers of the Greek philosopher Epicurus mocked Christian believers as childishly self-centered: "Christians are like a council of frogs in a pond, croaking at the top of their lungs, 'For our sake was the world created.'"

Many centuries later, Montaigne took a similarly scornful line. A *goose* was as entitled as any human to proclaim its eminence, Montaigne declared. "All the parts of the universe have me in view," he imagined his goose announcing. "The earth serves for me to walk on, the sun to give me light . . . I am the darling of nature."

But in the West, skeptics were few and scattered, and they stood little chance against the far more appealing doctrine that we humans are nature's darlings. Who would hesitate if the choice was *Are we special or are we geese?*

To deny our special status, and then to go even further and embrace the notion that many things are beyond our control, would be to ask a lot. It seldom happened.

Instead, scientists and theologians in the early 1800s took the idea that everything happened for a reason and worked it up into a full-fledged philosophy. This was a doctrine that fit with intuition, and people took it up with enthusiasm. Here was a way of thinking that could be captured in two short mottos—*Up with design! Down with chance!*—and both were hard to resist.

Physicists today talk about a "theory of everything." The doctrine that God had designed the world with humanity in mind was the nineteenth-century counterpart. It had the same grand sweep and psychological appeal, and it had the added advantage that it could be expressed in ordinary words rather than equations.

Some scientific theories have an improvised, duct-taped, Rube Gold-berg feel to them. Not this one. Here was an elegant, coherent way of embracing all the world's variety at once.

And then, in a small town in the south of England, the notion of "variety" took a spectacularly unexpected turn.

CHAPTER 10

Whales in the Treetops

Long before Mary Anning came along, people had collected fossils and stared at them with puzzled fascination. The appeal was easy to understand—these were rocks that looked uncannily like living things, like fish or ferns or leaves or seashells or bones.

But how could that be? How could a living creature turn to stone?

There were, in fact, *two* problems with fossils. First, they were made of the wrong thing—rock rather than bone or shell or wood. (The dinosaur "bones" in museums around the world are rock, not bone.) Second, they showed up in the wrong places—fossilized fish turned up inside solid rock; fossilized seashells turned up atop mountain peaks.

Everyone agreed that fossils *looked* as if they'd once been alive. But everyone knew, too, that rocks could take on strange and beautiful shapes. Think of crystals and geodes. They were indisputably lifeless but so intricate and eye-catching nonetheless that they seemed to show there were no limits to nature's ingenuity.*

---

* The term *fossil fuels*, which is centuries old, reflects the confusion about just what fossils were. The word *fossil* originally referred to *anything* unusual or valuable that had been dug from the ground.

Or look at the icy lacework that forms on windows on winter mornings. The names of the delicate patterns—they are sometimes called "fern frost" or "ice flowers"—suggest some organic secret behind the filigree. But those intricate networks have nothing to do with life.

Look a bit closer and matters grew even more perplexing. Could a clam, say, have journeyed hundreds of miles from the sea, crossed hills and valleys, then dug its way deep inside a cliff, and, after all that, cleared away all signs of its tunnel digging?

Don't be fooled by appearances, warned Martin Lister, a prominent English naturalist, in the late 1600s. Nature had the power to reshape lifeless bits of stone, deep inside the earth, into countless remarkable forms. But stone was stone. "There is no such thing as shell in these Resemblances of Shells."

Open a book by John Ray, a contemporary of Lister and perhaps the greatest naturalist of the era, and you can almost hear him yelping in frustration as he tried to make sense of the mysterious stones he had seen with his own eyes. Why was it, he asked, that all of the fossils he had examined resembled shells or clams or oysters? "Why should not Nature as well imitate the Horns, Hoofs, Teeth, or Bones of Land Animals, or the Fruits, Nuts, and Seeds of Plants?"

Ray had reason to be puzzled, because the formation of a fossil is a kind of slow-motion magic trick, as one substance transforms into another. Then, as if that weren't sorcery enough, the stony fish or shell turns up where it has no business belonging. *And now, sir, would you please open this sealed envelope and show the audience what you find inside!*

One of the first to sort out the true story was Robert Hooke, a bad-tempered, far-ranging English genius who did pioneering work, in the second half of the 1600s, in astronomy, mathematics, physics, and half a dozen fields besides. Hooke is far too little known, though, because he had the misfortune to make a lifelong enemy of Isaac Newton, who was even more brilliant and even less inclined to share the spotlight with a rival.

Hooke grew up on the Isle of Wight, in the English Channel. To this day the island's cliffs are rich with fossils, and as a schoolboy Hooke built a spectacular collection that he had dug up himself.

Beginning in around 1665, Hooke produced a series of detailed arguments in favor of the view that fossils were relics of living organisms. Fossils were *not* "apish Tricks of Nature," he snapped, nor were they striking but lifeless objects like geodes or crystals.

Just look at the objects that he and many others had unearthed. If nature were playing tricks, Hooke demanded, why did she always bury sturdy and solid objects, like teeth and shells, and never fragile ones, like flowers bursting with scent and color? "Why does she not imitate several other of her own Works? Why do we not dig out of mines everlasting vegetables, as grass, for instance, or roses?"

The answer was plain. "Bones, horns, teeth, and claws," as well as seashells, were "made of a stony substance which is not apt to corrupt and decay." *That* was why they endured through the ages. Bones and the other hard bits were old, not "newly generated" by some mysterious force of nature.

Hooke went further. Why did some fossils look distinctly different from present-day creatures? Hooke's answer was simple—the unfamiliar creatures were vestiges of extinct life-forms.

This was, at the time, a bold and controversial claim. In the eyes of biblical literalists, any mention of extinction was a rebuke of the divine plan, a slap in the Creator's face.

Hooke was a devout Christian, but he stood his ground. "There have been many other Species of Creatures in former Ages, of which we can find none at present," he declared forthrightly. Then he added another sentence, just as striking, that seems to anticipate evolution. "And 'tis not unlikely also but that there may be diverse new kinds now, which have not been from the beginning."

The story of life on Earth was a historical saga, Hooke argued, and the past had looked different from the present. Fossils were relics of those bygone days, akin to the coins found in ancient ruins. He cited the most revered of all vanished cultures. Fossils were "written in more legible

Characters than the Hieroglyphics of the ancient Egyptians, and on more lasting monuments than those of their vast Pyramids and Obelisks."

Hooke was a skilled artist as well as a scientist. This engraving is based on his drawings of fossils that he found on the Isle of Wight.

But how had living materials like wood or bone changed into stone? Hooke had seen that answer, too. It was a hunk of petrified wood, "a piece about the bigness of a man's hand," that gave him the crucial clues.

The key was that, over the course of eons, "*petrifying* water (that is, such a water as is well *impregnated* with stony and earthy particles)," had seeped into every nook and cranny of a buried piece of wood.

The language was old-fashioned, but the concept was modern, and Hooke emphasized that it applied to fossils generally and not just to petrified wood. Mineral-rich water insinuated its way into a piece of wood or a fern or a fish skeleton. When the water dried, the minerals remained behind. In denser objects, like shells, a similar but even slower-paced process took place. If that transformation happened slowly and gently enough—the process might take thousands or even millions of years—you ended up with a stony duplicate of what had once been a living organism.

That explanation was far ahead of its time. But Hooke immediately moved on to the even more difficult question of why fossils turned up in such unlikely places.

By Hooke's day, this was already a long-standing mystery. The question had tantalized naturalists for so long that it had given rise to a sort of emblem—a seashell on top of a mountain—that served as shorthand for the whole vexed subject.

Leonardo da Vinci had tackled the mystery long before, in the early 1500s. On his excursions into the countryside around Vinci, in Tuscany, he had climbed mountains and found seashells near the summit. He deduced that land and sea had changed places. This was not quite seeing the world in a grain of sand, perhaps, but it was close.

Leonardo was absolutely correct, but, even centuries later, his insight could make the most level-headed scientists dizzy. Whales and walruses now swam in the very spot, marveled Thomas Huxley in 1868, "where birds had twittered among the topmost twigs of the fir-trees."

But in Leonardo's day, and for long afterward, conventional wisdom took a different tack. The reason that fossilized fish and seashells sometimes turned up far from the sea, everyone agreed, was that they had been left behind by the biblical flood.

That was a "foolish and simpleminded" notion, Leonardo grumbled in his notebooks. Dig in a quarry or probe an exposed cliffside, he wrote, and you found fossils at different layers in the rock. How could one flood explain that? Leonardo had a better idea. "From time to time," he wrote in his notes, "the bottom of the sea was raised, depositing these shells in layers."

But Leonardo, who sometimes seems as much a time traveler as the fossils he studied so intently, never published a word about any of these ideas. They were left for later scientists, like Hooke, to rediscover.

Hooke's tone was often grumpy and impatient—a seashell on top of a mountain was a fact and not a mirage—and he complained that he had spent too much time considering the "imaginary influences" that might have set it in place. (Perhaps, he sniped, it had been "placed there by Merlin.") The seashell was evidence that needed to be explained, like "an anchor found at the top of a hill, or an urn or coins buried underground."

To Hooke's mind, the explanation was easy to see. The seashells were "a very cogent argument that the superficial parts of the earth have been very much changed since the beginning, that the tops of the mountains have been under water, and consequently also that diverse parts of the sea's bottom have been heretofore mountains."

That was Leonardo all over again, and even in Hooke's day it was still impressively ahead of its time (though Hooke's explanation of what had lifted the seas and drowned the mountains—earthquakes—proved incorrect). What was perhaps less impressive is that no one paid much attention.

Deep into the 1700s, the whole subject of fossils continued to make people uneasy. The notion of seas and mountains rising and falling like toddlers on a seesaw was especially troubling. Even Voltaire, an ardent admirer of science (and a fierce enemy of organized religion and arguments that invoked God), rejected it.

Unadorned common sense could easily explain how fossilized shells and fish might have come to rest atop mountains, Voltaire wrote, but learned philosophers insisted on dreaming up far-fetched explanations.

He had a better and far more plausible explanation—picnickers. "It is much more natural to suppose that these fish had been brought thither by some traveller, who, finding them spoiled, threw them away, and, in process of time they became petrified."

Or perhaps the fossils were souvenirs that had been misplaced by Crusaders headed home from the Holy Land? "Why may it not be remembered, that the innumerable crowds of pilgrims and Crusaders . . . brought back with them a number of shells?"

CHAPTER 11

Without a Trace

By Mary Anning's day, the debate was all but over. With the exception of a few religious holdouts, everyone agreed that fossils were relics of ancient plants and animals, formed by ever-so-slow natural processes.

They weren't stones that, by coincidence, happened to look like fish or ferns (as clouds sometimes look like camels or castles); they weren't objets that God had made for his own amusement; they weren't traps that Satan had planted to sow confusion among believers.

But questions remained. Why was it, for instance, that fossils were so rare?

Today, centuries later, it seems that we finally see the answer. If your ambition was to become a fossil, think of all the obstacles you'd have to overcome. You couldn't die just anywhere, for starters.

Dying at sea would be your best option. At least then you'd have a chance of drifting down to a quiet grave on the ocean floor, where sand and mud might quickly bury you. But not just any sea would do.

The shallow seas near a coral reef, say, would be a bad choice, because they teem with grave robbers. Crabs and starfish and shrimp chow down on dead bodies, and bacteria finish the job. Near reefs, bodies can vanish without a trace. The oxygen-poor depths of the ocean, where there are fewer microorganisms or scavengers, would be a better bet.

Die on land and your prospects would be immensely worse. A corpse in the open air quickly draws a raiding party of scavengers—jackals, hyenas, coyotes, ravens, vultures; and an assortment of beetles, ants, and flies; and colonies of bacteria—that tear into flesh and devour bones.

Wind and weather make a bad situation worse. If a skeleton somehow manages to stay intact, floods and storms might break it apart or scatter the bits across a vast territory.

The upshot is that the huge majority of fossils—99 percent of all the fossils ever found—come from sea creatures like sharks and shellfish. But the other 1 percent includes many of the creatures we care about the most.

Dinosaurs were the most conspicuous one-percenters. They lived on land, which means that we've lost nearly all evidence that they ever lived at all. Out of every eighty million Tyrannosaurus rexes, scientists calculate, *only one* was ever fossilized. (The total number of T. rexes in museums around the world is around three dozen.)

When dinosaur fossils do turn up, it's nearly always as the result of some fortunate fluke. Take stegosaurus. It is a star among dinosaurs, as we've seen, but fewer than one hundred stegosaurus skeletons have ever been found. Each of those finds was a needle in a field of haystacks. One of the best finds is now on display at the Natural History Museum in London. By good fortune, the stegosaurus bones—some vertebrae, a hipbone, a thigh bone—happened to wash into a river in prehistoric Britain. The river carried them to a shallow sea where they lay buried for 150 million years. The bones turned up in 1873, near Oxford, when the Swindon Brick and Tile Company accidentally unearthed them.

Few fossils survive the eons intact. A bone or a fragment of bone or a jumble of bones is far more common than a complete skeleton. But once or twice an eternity, a fossil is entombed so suddenly—this is akin to the insect-trapped-in-amber scenario from *Jurassic Park*, except on a colossally larger scale—that it provides a snapshot of a moment from tens of millions of years ago.

In the Gobi Desert, in Mongolia, two dinosaurs fighting a death match were instantly buried when a sand dune collapsed on top of them.

The skeletons are still entwined; the velociraptor (on the ground, in the drawing below) has sunk its claw into the protoceratops's skull, and the protoceratops has its jaws clamped on the velociraptor's leg. The battle took place seventy million years ago.

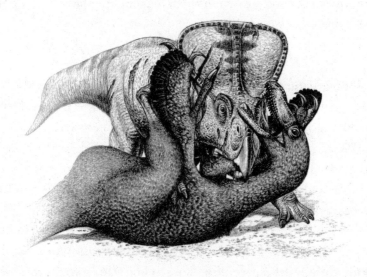

Not every snapshot is gory. Also in the Gobi Desert, and also about seventy or eighty million years ago, a sandstorm trapped a feathered dinosaur, citipati, as it sat in its nest keeping its eggs warm.

In comparison with fossils from land creatures, fossils from sea dwellers can seem positively abundant. Today, in Charmouth, England, a mile or two from Lyme Regis, a plastic crate sits outside the front door of a fossil shop. It is full of fragments of ammonites—worn and incomplete but easily recognizable—with a handwritten label, *Ammonite bits, 175 million years old, £2 each.*

They come from the field of a farmer who lives nearby. When he finishes plowing, a local fossil hunter wanders through the turned-up soil and gathers them up.

Even for animals that lived at sea, that sort of bounty is rare. But it does happen. In the same corner of the world as the ammonite field—this is very near where Mary Anning made her finds two centuries ago—professional collectors tell stories of the 1970s, a flukishly golden era for excavating ichthyosaurs. As collectors struggled along the beach, carrying their prize, they sometimes spotted *another* ichthyosaur peeking out from the bottom of a cliff, where pounding waves had exposed long-hidden rocks.

But the big picture is far less rosy. Whether you die at sea or on land, the odds that you will become a fossil are staggeringly low. The problem is that, even if you do manage to escape beetles and bacteria and starfish and other scavengers, your trials have only begun. For Earth's crust is a violent, dynamic place where countless tons of rock are perpetually on the move, and a would-be fossil's next task is to withstand a gauntlet of squeezing, melting, twisting, stretching, and grinding.

It would be as if a swimmer in a capsized boat had to make it to shore past whirlpools, crashing waves, and colliding currents. Worse, actually. For the analogy to hold, some of those roiling waters would be burning hot, for instance, and the ordeal would continue for eon after eon.

And after all that, you would still need to be found. Not just found, in fact, but found quickly.

"Whenever there's a storm, we rush out right away," says Phil Davidson, an English fossil hunter with a long string of discoveries to his credit. "It's a race against time, because the sea uncovers the fossils but it can also damage and destroy them, maybe with the next tide. There's a fine line between discovery and destruction.

"Many fossils are already damaged," Davidson continues. "That can be the clue that there's something hidden inside a rock. Maybe it's the edge of an ammonite peeking out or the tip of an ichthyosaur jaw. These things have been waiting for hundreds of millions of years, but now you need to hurry, and the clock is against you."

If you *do* get to the bones in time, your troubles may still not be over. On an expedition to England's southern coast in 1828, Gideon Mantell, one of the nineteenth century's renowned fossil hunters, found "an enormous bone weathering out of the chalk cliffs." Mantell set to work. (His name is pronounced "mantle.") "After three hours hard labor," he wrote, "I succeeded in laying bare a bone 30 inches in circumference and nine feet long."

He lifted his treasure free, but "in attempting to remove it it fell into a hundred pieces! A few fragments were the only relics I could bring away of this, the most magnificent fossil I ever discovered."

It is a cautionary tale, and historians retell similar stories with an empathetic shudder. "The old horror movie trope of a long-hidden skeleton turning to dust at the first touch," one expert writes sadly, "has a basis in reality."

And even then—even after having dodged the scavengers, and survived the geological roller coaster, and weathered the ravages of time, and won the lottery of discovery—even then you could not relax. Even tucked away inside a museum drawer, you might not be safe.

Many fossils contain pyrite, "fool's gold," and pyrite can combine with oxygen and water in the air to form a yellow-gray "rust" and sulfuric acid. The acid can reduce a fossil to powder. Davidson, the English fossil hunter, can scarcely contemplate the subject without wincing. "There are countless stories of specimens being left in drawers for years and then opened again to find out they have turned to dust!"

Under assault by acid, the fossil itself can disintegrate. So can the tray holding the fossil, or the wooden drawer holding the tray, or even the fossils in a different drawer.

The bottom line is stark. Ninety-nine percent of all the animal species that ever lived—not individuals but *species*—leave no trace whatever.

The biologist Jerry Coyne once observed that paleontology is one of the few fields—theology is another—"in which the students far outnumber the objects of study."

CHAPTER 12

"None of the Advantages"

Mary Anning was "arguably the first person on the planet to take up fossil hunting as a full-time career and occupation," in the judgment of one modern historian. It was not a career likely to draw many candidates.

We have seen already that it was both low paying and dangerous. "This persevering female," one of her fellow collectors wrote admiringly, "has for years gone daily in search of fossil remains of importance at every tide, for many miles under the hanging cliffs at Lyme . . . at the continual risk of being crushed."

The risk was real. One October day, in 1833, a crumbling cliff brought tons of rocks crashing down next to Anning. As always she had her dog, Tray, with her. "The Cliff fell upon me and killed him in a moment before my eyes and close to my feet," she told a friend. "It was but a moment between me and the same fate."

For fossil collectors at Lyme, the risk was part of the routine. Richard Anning died in a fall from a cliff, and he'd nearly been killed in rock slides before that. There was nothing to be done but shrug it off and head out again. Bad conditions meant good hunting.

Mary Anning worked relentlessly, setting out each morning in the dark and cold and then searching and digging hour after hour. ("For the last year

I have been very unsuccessful," she wrote to a friend in the autumn of 1833, "but hope as the stormy season is coming on I may be less so.") Even when she was so ill that she had to be "brought fainting from the beach," as one friend put it, she refused to let up. There were no fossils to be found at home.

When she did see something promising, on the other hand, it was easy to get too caught up to heed danger signs. Anning spotted a plesiosaur bone one day. The tide was coming in fast, but Anning's attention was elsewhere.

"I was so intent on getting it out that I had like to have been drowned," she told a friend soon afterward, and she was not exaggerating. Pummeled by the waves and soaking wet, she asked a man who had been helping her why he hadn't warned her about the tide. "He said he was ashamed to say I was frightened when you didn't regard it."

On most days, Anning dodged the waves with the nonchalance born of experience, in much the way that pedestrians racing to work today pick their way between buses and taxis at crowded intersections.

"In one place we had to make haste to pass between the dashing of two waves," reported a woman who joined Anning on a morning trip to the beach. "Before I knew what she meant to do, she caught me with one arm round the waist and carried me for some distance, with the same ease as you would a baby."

But the savviest jaywalker can get flattened by a bus.

Anning's coolness grew partly from experience. She had lived with stormy seas and crumbling cliffs from childhood on. At age sixteen she had found a woman's drowned body on the beach. A ship called the *Alexander*, on its way from India to London, had capsized in a hurricane, and 130 people had perished. One of the victims washed ashore near Lyme. "Mary untangled the seaweed which had attached itself to her long hair and performed all the other offices due from the living to the dead," a friend named Anna Maria Pinney wrote in her journal, "and the unknown corpse being deposited in the Church until some friend appeared to claim it, she daily went to strew fresh flowers over it."

Life on land held its own dangers. Two months after the rockfall that killed her dog, Anning set out for another day's collecting. While she was still in town, a runaway cart on a narrow bridge knocked her down. "Yesterday," her friend Elizabeth Philpot wrote in a letter, "she had one of her miraculous escapes in going to the beach before sunrise and was nearly killed in passing over the bridge by the wheel of a cart which threw her down and crushed her against the wall."

Anning had always been deeply religious, and these reprieves from death strengthened her faith. "The Word of God is becoming precious to her, after her late accident," Anna Maria Pinney wrote in her journal, although it is not certain just which accident she had in mind.

For the most part, though, Anning's faith showed itself more in practice than in theory, more in helping the needy and scrutinizing the natural world than in poring over biblical texts. She honored God's works by studying them.

Mary Anning, with her geological hammer, in a portrait from 1842

Still, there was an irony built into Anning's fossil hunting. "Her faith let her do this dangerous work—and it *was* dangerous work," the historian Thomas Goodhue observes. "She survived several very close calls with death. She did this work because of her faith, and the things that she found upset the faith of millions of people, across the nation and around the world."

Mary Anning faced down storms and other natural dangers, but she had to combat a host of man-made hazards, as well. Nearly all the geologists and fossil collectors of her era were well-to-do. Anning was not.

Her collector friend Henry De la Beche had inherited a sugar plantation in Jamaica (and the enslaved people who worked it). The collector Thomas Hawkins had inherited a fortune, too, and promptly squandered it on fossils. (By the time he went broke, his collection weighed twenty tons.) Her friend Charlotte Murchison and Charlotte's husband, Roderick, both geologists, spent the first two years of their marriage sightseeing their way across Europe.

Anning had no such options, and any sort of education beyond the rudimentary was out of the question, too. Anning went to school for only three years. By about age eleven her schooldays were behind her.

"She was the right person in the right place at the right time, but she was the wrong sex," the historian Jo Draper commented recently. The observation is true, as far as it goes, but it may *understate* the obstacles that lay in Anning's path.

She was the wrong class, as well, which meant that she was routinely overlooked and undervalued. It meant, in addition, that Anning's fossil hunting had an urgency to it that the better-off never knew.

As an adult, and in spite of her lack of schooling, she steeped herself in the scientific literature. Keeping up to date was harder than it sounds because technical journals were expensive. Some of Anning's scientist friends sent her printed copies of their papers, but others forgot (even when those papers discussed fossils that Anning had found).

Anning resorted to copying articles by hand—she copied the illustrations as well as the text—and she drew even the tiniest bones so carefully that the copy and original are often nearly indistinguishable.

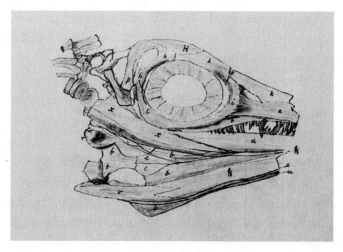

Ichthyosaur drawing by Mary Anning

It was tedious work. (She took a break in the middle of copying one long paper to sketch her dog Tray, who was, as usual, at her feet.) Long-winded authors made the task worse. Anning finished copying one long paper by Conybeare (on ichthyosaurs and plesiosaurs, her prize discoveries) and wrote, exasperatedly, on the last page, "When I write a paper there shall not be but one preface."

She never got the chance.

Frustrations came on other fronts, too. Museums happily put Anning's fossils on display, for instance, and they made a point of citing the name of the donor responsible for those handsome gifts. Mary Anning's name went unrecorded.

She was perpetually worried about money, especially because she had to support her elderly mother as well as herself. (Even so, Anning had a reputation for generosity. She had "more than the power of an eagle's eye" for finding fossils, one friend marveled, but the eagle eye could become a blind eye when necessary. "She were good to the poor, she were," one Lyme Regis local recalled. "Whenever she found a little keg on the shore [containing smuggled whiskey or other contraband], she would cover it up, and not let the preventative men see it, but would tell some poor person of it.")

For long stretches Anning could not find enough fossils to keep herself afloat—"therefore it is my intention to look out for some other pursuit," she lamented in 1826—and when she did find buyers, she had to chase down payments they had neglected.

She sent polite letters about overdue bills, but the urgency shone through. "Mary Anning takes the liberty of reminding Mr. Sowerby that the two months expired on the first of May," she began one message in 1829. She ended on a plaintive note (using the word *wants* in its old sense of "needs"). She wrote, she explained, because "she really wants the money."

Anning knew her own worth, and perhaps it grew tiresome to have to fight for what was only her due. "She says the world has used her ill," one friend recalled. "These men of learning have sucked her brains and made a great deal of publishing works, of which she furnished the contents, while she derived none of the advantages."

Despite the snubs and the money woes, Anning's reputation was secure from her midtwenties on. With ichthyosaurus and plesiosaurus to her credit, geologists and fossil collectors all knew her. (So important were those finds that the Geological Society considered a design for its coat of arms that featured an ichthyosaur on the left and a plesiosaur on the right.)

Anning would continue on for another two decades, racking up more coups along the way. In 1828, for instance, she found the first flying reptile ever discovered in Britain, a raven-sized creature called a pterosaur that looked a bit like a bat.

William Buckland, the geologist who had earlier tried to describe plesiosaurus, searched again for a useful comparison. He mentioned bats and crocodiles and iguanas, and then gave up. "In short, a monster resembling nothing that has ever been seen or heard of upon the earth."

Anning had already discovered two creatures that had roamed ancient seas. Now she had added a denizen of the sky.

Eminent scientists sought her out, and most—with notable exceptions, like Conybeare—happily acknowledged not just her finds but also her insights. She could glance at an isolated fossil bone, for instance, and identify the species it belonged to, or see how a jumbled array of bones fit together.

Buckland, one of the most prominent geologists of the era, led the parade of admirers. He and Anning were lifelong friends—they hunted for fossils together, wading in knee-deep water and probing the cliffs even as the tide rolled in—and Buckland seized every opportunity to praise "the industry and skill of Miss Anning" and to point out that "Miss Anning informs" him of this, and "Miss Anning has observed" the other.

Buckland noted approvingly, for instance, that Anning had carefully dissected modern-day squid to see how their anatomy compared with that of squidlike fossils. Together with a close friend, the dauntingly talented Elizabeth Philpot, she even managed to resurrect millions-of-years-old squid ink.

Elizabeth had moved to Lyme Regis with two of her sisters in 1805. All three sisters became avid fossil collectors, though Elizabeth took the lead and quickly gained a reputation as an authority, especially on fish fossils. She was, as well, a skilled artist.

This is a drawing of an ichthyosaur's head, by Elizabeth Philpot.
She painted the skull using ink from a fossil squid (she had devised
a technique for reviving the ink by mixing it with water). The ink,
like the ichthyosaur, was two hundred million years old.

In letters, books, and lectures, other prominent scientists added their own tributes to "the celebrated Miss Anning" and "the Princess of Paleontology." Soon newspapers and magazines spread the word, and before long curious visitors of all sorts found their way to Anning's Fossil Depot at Lyme Regis.

One aristocratic visitor from London noted with astonishment, when Anning was only twenty-five, that "this poor, ignorant girl" was "in the habit of writing and talking with professors and other clever men on the subject [of fossils], and they all acknowledge that she understands more of the science than anyone else in this kingdom."

CHAPTER 13

"Sister of the Above"

A few of the visitors who made pilgrimages to visit Mary Anning declared themselves unimpressed. In 1832 the physician and fossil hunter Gideon Mantell "sallied out [to Lyme Regis] in quest of Mary Anning, the geological lioness . . . ," as he recalled later. "We found her in a little dirty shop, with hundreds of specimens piled around her in the greatest disorder."

The hunter and the lioness seem to have eyed each other warily. Anning left no record of the encounter, but Mantell noted—perhaps with surprise—that she had a sharp tongue and a keen sense of her own merits. "She, the presiding deity, [proved] a prim, pedantic, vinegar-looking, thin female," Mantell noted, "shrewd, and rather satirical in her conversation."

It is hard to know whether Mantell's picture reflects Anning's sourness or his own. Anning never married—she was, in nineteenth-century terms, a "spinster"—and Mantell's description of her as a grim, prim scold may be less a portrait than a bit of off-the-shelf derision.

Anning's "satirical" bent, too, may have had mostly to do with impatience and self-respect. Anning kept a notebook where she copied down poems and passages from her reading (books and magazines, like scientific journals, were too expensive to buy). One long entry was headed

"Woman!" and began, "And what is a woman? Was she not made of the same flesh and blood as lordly Man? Yes, and was destined doubtless, to become his friend, his helpmate on his pilgrimage but surely not his slave, for is not reason hers?"

There is no doubt Anning was outspoken, and this in an era when women, and working-class women especially, were expected to defer to their "betters." Anning had little time for deference. One astonished friend, only nineteen years old at the time, marveled that Anning had been "noticed by all the cleverest men in England, who have her to stay at their houses and correspond with her in geology, etc." Not only that: "She glories in being afraid of no one," Anna Maria Pinney went on, "and in saying everything she pleases."

Anning had a host of loyal friends, and they helped her through hard times. As early as 1820, when Anning was not yet twenty, a fossil collector named Thomas Birch sent a note to Gideon Mantell—he of the "vinegar-looking" remark—explaining why he had nothing to sell him. "The fact is that I am going to sell my collection for the benefit of that poor woman [Mary's mother] and her son and daughter at Lyme who have in truth found almost *all* the fine things, which have been submitted to scientific investigation."

Birch swallowed hard—"I may never again possess what I am about to part with"—but he stuck to his vow and put his collection up for auction. (This was a doubly good deed because Birch had purchased many of his prize specimens from Anning in the first place.) The sale brought in £400. Birch handed nearly all of it to the Annings.

That was a considerable sum, as much money as a craftsman could earn in several years. By way of comparison, Anning sold her best ichthyosaur skeleton for £23 (about half a year's wages for a skilled workman) and her best plesiosaur for £100.

In 1829, Anning wrote to her friend Buckland about her latest fossil find. She had discovered another plesiosaur, which today is at London's Natural History Museum. "How I wish you could see it are you likely to visit Lyme soone," she wrote in an excited, unpunctuated rush, and she reported her disappointment that the museum in nearby Bristol had offered her only £30.

In 2020, Sotheby's put Anning's letter—*not* the plesiosaur itself—up for sale. The winning bidder paid £100,800.

In 1830, when Anning was pinched for money once again, William Buckland and the geologist Henry De la Beche cooked up a face-saving scheme to provide money to their old friend. De la Beche was an artist as well as a scientist. He made a print that depicted Anning's most important finds in action—ichthyosaurs and plesiosaurs tangled in mortal combat in the sea while pterosaurs careened in the sky.

Devon in Ancient Times, as the image was called, knocked viewers sideways. Static drawings of bones were old news. This was different. "Exuberant, chaotic, and revolutionary," in the words of the historian Zoë Lescaze, it was the first depiction of animals in a world before humans.

Countless drawings of the prehistoric world have followed in its wake. That flood of images makes it hard to imagine the shock of the original. But *Fantasia* and *Jurassic Park* and all their elaborate offspring are twigs on a family tree that has a watercolor by Henry De la Beche at its base.

No one knows how many prints were sold, but the proceeds were considerable. (The prints went for two pounds and ten shillings each, about a month's wages for a laborer.) All the money went to Anning. But no matter what, she seemed destined for trouble, financially speaking.

By the early 1830s she had finally managed to acquire a nest egg. It didn't last. In 1833, according to newspaper reports, "Mary Anning, a female in humble life residing at Lyme Regis," lost all her money to what sounds suspiciously like a con man. She had turned over her life savings, £200, to "a gentleman whom she had entrusted, without receiving any acknowledgment" (i.e., without a receipt).

Soon the money (and the gentleman) were gone, the newspaper went on, and Anning was left with "the distress of mind consequent upon her loss."

Buckland came to the rescue once more, persuading his fellow geologists to kick in to replace Anning's lost money. Then he and several other members of the Geological Society convinced the government to grant her a pension.

"It wasn't a fortune," notes Tom Sharpe, Anning's best biographer, "but it provided Mary Anning with a modest annual income of £25, approximately twice that of a housemaid." That modest subsidy, historians suggest, made Anning the first woman to receive government support for her scientific work.

A geologist friend of Anning's painted this watercolor and then sold prints to raise money for her. It shows the first creature Anning discovered attacking the second creature she discovered, while flying creatures—her third major discovery—soar overhead.

Mary Anning, who died young, spent her brief life digging up the bits and pieces that were the last relics of vanished lives. Perhaps it is fitting that she, too, has largely vanished from history.

What was left, it seemed for many years, was mainly one easily overlooked scrap from the past. That relic was not bone or tooth but a few mumbled words of not-quite-poetry. For decades, geologists and historians have written that Anning was the inspiration for "She sells seashells by the seashore."

The story turns up repeatedly in textbooks, histories, and biographies as a bright note in a dark story. (The claim even shows up twice, in prominent spots, in London's Natural History Museum.) Alas, it turns out not to be so.

"She sells seashells" was apparently dreamed up by elocution teachers, in the mid-1800s, as a kind of vocal workout. It had nothing to do with Lyme Regis's too-little-heralded fossil finder. Passed over during her lifetime, it seems that Mary Anning was not even granted the dubious honor, after her death, of commemoration in a tongue twister.

In recent years, though, she may finally have begun receiving proper recognition. In 2010, the Royal Society, the world's best-known scientific organization, put together a list of the ten most influential women in the history of British science. Anning turns up there, along with such notables as Rosalind Franklin, of DNA fame.

In 2018 the Bank of England announced its decision to put a scientist on the £50 note. A list of nearly a thousand names was whittled down to a dozen, and Anning made the short list. Fittingly, perhaps, for a woman so chronically short of money, the bank passed her over in the end (in favor of Alan Turing).

Anning spent her entire life in the small town of Lyme. (Her home no longer exists, but the Lyme Regis Museum, which houses a small and meticulously curated collection, stands on the same site.) It seems she left only once, at age thirty, when her friend Charlotte Murchison invited her to London for a visit.

"I have never been out of the smoke of Lyme," Anning told Murchison in a letter, and she happily accepted the offer. While in London she visited the Geological Society and the British Museum (curiously, her notes mention such wondrous sights as Egyptian mummies but make no reference to fossils).

In her short life Mary Anning had discovered a new world, but this journey of 150 miles would be the farthest she ever ventured from home.

She died at forty-seven of breast cancer. (The disease was virtually a death sentence in her day. In the notebook in which Anning copied down passages from her reading, she included one brief entry: "Affliction . . . brings us the nearest way to God.")

Her grave lies in a small churchyard only a few minutes' walk from her home. But her name is not the first you see on her gravestone. "Sacred to the memory of Joseph Anning, Who died July the 5th, 1849, Aged 53 years," the inscription begins.

Next come a few sad, simple words: "Also of three Children who died in their Infancy."

Then come the final lines, which seem dismayingly like an afterthought. "Also of Mary Anning, sister of the above, Who died March the 9th, 1847, Aged 47 years."

The grave perches at the edge of the sea, nestled near the cliffs that Anning knew so well. Visitors leave flowers and seashells, and, sometimes, a toy dinosaur or two.

CHAPTER 14

Ferns and Fox Hunters

On the same February evening in 1824 when William Conybeare told the Geological Society about Mary Anning's bizarre, long-necked plesiosaur, William Buckland dropped another bombshell.

Both speakers on that remarkable night were scientists and devout Christians; both were, in fact, clergymen. Buckland had newly been elected president of the Geological Society. This was his first meeting in his new role. As lively a speaker as Conybeare was dreary, he came as near as anyone could to pulling back a curtain and unveiling a house-sized, teeth-baring beast.

Buckland's talk had been a long time in the making. Six years previously, in 1818, he'd had a visit from Georges Cuvier, a towering figure in French science who was often hailed as "the father of paleontology." Buckland was an Oxford geologist and an eminent scientist in his own right.

The two men had set to work trying to make sense of a jumble of prehistoric bones at Oxford's Ashmolean Museum. One bone was especially conspicuous. This was a huge femur that looked exactly like Robert Plot's drawing of a pair of fossilized testicles.

Buckland and Cuvier proposed that the giant bone belonged with a jawbone (complete with teeth) and with fragments of pelvis and shoulder. What had been dismissed as a jumble of odds and ends in Oxford's collection, the two men suggested, all came from a few closely related individuals.

The tooth-studded jawbone was the most valuable clue, because there was no question that jaw and teeth belonged together, and it was clear besides that these were reptilian teeth. So much for Plot's belief that he had found a bone from a giant or an elephant (or any other mammal).

Just showing that the scattered bits came from the same *era* would have been a coup; to show that they all belonged to an immense and hitherto-unknown animal was an enormous triumph.

Now, at this February 1824 meeting, Buckland was ready to tell the world what he had figured out. He did not have anything like a complete skeleton to show his fellow geologists, but he had seen enough bones and teeth for a mental reconstruction. He painted a garish picture for his enthralled audience.

The creature had sharp teeth six inches long. It was a meat eater and a reptile (teeth are the giveaway: unlike mammals' teeth, reptiles' teeth are all the same shape, although different in size). But this bizarre reptile had stood upright like a mammal rather than on splayed-out legs like other reptiles. It was enormous, besides, more than twice as long as a crocodile and twice as massive as a rhinoceros.

Buckland suggested that it be named megalosaurus. The name was a hybrid of two Greek words, *mega*, meaning "giant" (as in *megaton* or *megalomaniac*), and *saurus*, meaning "lizard." "Giant lizard," then, a fitting name for a creature perhaps forty feet long and as massive as an elephant.

Megalosaurus had an odd, hybrid look, as if it had been assembled by a sculptor in a hurry. Its four sturdy legs might have belonged to an oversized rhinoceros; its head seemed suited to a crocodile.

The unlikely beast would later be hailed as the first dinosaur ever identified, though in 1824 the notion of "dinosaur" was still more than a decade off.

Dickens nonchalantly placed a megalosaurus into the opening scene of *Bleak House*, in 1852. "London," he began. ". . . Implacable November weather. As much mud in the streets as if the waters had but newly retired from the face of the earth, and it would not be wonderful [i.e., astonishing] to meet a Megalosaurus, forty feet long or so, waddling like an elephantine lizard up Holborn Hill."

A megalosaur (and companion) as they were pictured in the 1800s, drawn by the artist and sculptor Benjamin Waterhouse Hawkins

This was, the writer Kathryn Schulz observes, "the first sighting of such a creature in England in a hundred and sixty-six million years, and the weirdest walk-on in all of Victorian literature."

"Waddling" was a straight lift from Dickens's scientist friends, but it seems to have been a mistake. Scientists today have set megalosaurs and many other dinosaurs on their hind legs and raised them up into the air. Many were built for speed. Tyrannosaurus rexes, for instance, seem to have been frighteningly nimble, especially when they were young. Juvenile T rexes were, in the words of one starstruck paleontologist, "ballerinas of doom."

Buckland did not quite shout "Run!" on the night in 1824 when he provided the first description of his megalosaurus, but he lingered happily on all the frightening details. And he told his listeners that the fossil hunter Gideon Mantell, who was in the audience that evening, had found an even larger specimen! Mantell's "great fossil lizard," Buckland suggested, had measured perhaps sixty or seventy feet!

Megalosaur as it is currently imagined

Buckland's guess turned out to be wrong. Mantell had found not another megalosaur but a different creature altogether. It would take a few more months for the story to come into focus, but in one of paleontology's great advances and with a few strange teeth as crucial clues, Mantell would soon prove that England had once been home to yet another giant, extinct reptile.

Mantell's discovery was especially noteworthy, for it was simultaneously a triumph of imagination and a tribute to persistence. Gideon Mantell was a handsome, charming country doctor who had been obsessed with fossils since childhood. Plagued through much of his life by spectacular bad luck (and spectacularly bad decision-making, especially when it came to spending money on fossils), he floundered from crisis to crisis, always convinced that he was one discovery away from winning the scientific esteem he craved.

It was not to be. Snobbery played a large role in the story. Mantell was intelligent and ambitious, but he was an outsider striving to make his way in a rigidly class-conscious society. Scientific organizations at the time were essentially gentlemen's clubs. Pedigree often counted for more than talent.

The geologist Roderick Murchison, to take one example of many, had never given a thought to science until his wife convinced him that he was frittering away his life. (This was Mary Anning's geologist friend Charlotte Murchison.) "The noble science of fox-hunting was then my dominant passion," Roderick recalled many years later, but Charlotte convinced him to give geology a try.

Just two years after his change of course, Murchison was elected a fellow of the Royal Society. Murchison's qualifications, as the president of the society put it, were that "he was an independent gentleman having a taste for science, with plenty of time and enough of money to gratify it."

Mantell, a shoemaker's son, found no such welcome. He forged ahead even so, working fanatically to balance a more–than–full-time medical practice with endless forays in search of fossils. His fossil collection would eventually outgrow his house, ruin his career, and destroy his marriage (he collected so many fossils that he literally turned his house into a museum).

But before it all went wrong, it went fabulously right.

All the early scientists and fossil hunters knew one another (it was Mantell who'd had a testy encounter with the "geological lioness," Mary Anning). Mantell and Anning had different home territories, though, which meant that they happened on different fossils.

Mantell lived in Sussex, which was south of London and about 150 miles east of Mary Anning in Lyme Regis. The landscape of Sussex in ancient times, Mantell soon found, seemed strangely unlike that of ancient Lyme Regis.

Along with giant, fossilized bones, Mantell had begun to uncover giant, fossilized palm trees and ferns. The thought of those tropical interlopers took some getting used to. Britain was famously the home of sturdy oaks and stately elms, not towering palms and twisting vines that thrived in year-round sun and steamy heat.

But Mantell's depiction of lush tropical landscapes was not entirely unfamiliar to English readers. In 1771 the great explorer James Cook had brought home hundreds of tropical plants and trees. Cook would make three epic voyages altogether—he would visit Tahiti and Australia and New Zealand and Hawaii—and the exotic tales that he and other explorers brought home proved as enticing as any of the plants and seeds they carried in their cargo holds.

Europeans of Cook's era, and for another generation or two after that, found the sheer exuberance of life in the tropics almost impossible to fathom. "We are here in a divine country," the scientist and explorer Alexander von Humboldt exulted, in 1799, when he first stepped ashore in Venezuela. In the tropics everything was huge and new and bursting with life.

"We rush around like the demented," Humboldt babbled in a letter to his brother. "In the first three days we were quite unable to classify anything; we pick up one object to throw it away for the next." Humboldt and a botanist companion staggered from one discovery to another. "Bonpland keeps telling me that he will go mad if the wonders do not cease soon."

That there could be such things as vines as thick as cables and plants with leaves the size of serving platters was hard to fathom. That they could once have been where they did not belong was nearly impossible to take in. Could cold, gray England really have been home to tropical marvels?

Mantell never traveled to the tropics and explored for himself, like Humboldt, but he slowly pieced together a picture of England's remote, sultry past. Unlike the ancient sea that Mary Anning had found, Mantell had begun to uncover an ancient landscape—emphasis on *land*—where palm trees stood tall on the shores of rivers, and ferns blanketed the ground. He pictured "a mighty river flowing in a tropical climate over sandstone rocks . . . through a country clothed with palms and ferns . . . inhabited by turtles, crocodiles, and other amphibious reptiles."

Such descriptions of once-upon-a-time landscapes are commonplace today, but Gideon Mantell, writing in the 1820s, had no precedents to draw on. And his scenario involved not only an unfamiliar world but also a world inhabited by enormous, plant-eating reptiles. No one had heard of such creatures, and no one listened.

Mantell's Exhibit A was a collection of large, odd-looking fossil teeth. They were big, and they had been found along with huge bones, so whatever creatures they came from had been immense, too.

What was puzzling was that they didn't look like the teeth of a carnivorous reptile, which are sharp, pointy cones well suited to tearing into chunks of meat. (When reptiles close their mouths, their teeth don't meet; they gulp down their food in big pieces rather than chew it into little bits. Mammals employ a different strategy, taking advantage of what the naturalist David Rains Wallace calls "a Swiss army knife of incisors, canines, premolars, and molars.")

Mantell's fossil teeth looked more like a herbivore's, which are broader and flatter, the better to grind up plants and grasses.

Why was that strange? Because the teeth had been deeply buried under Tilgate Forest, which Mantell knew so well. That seemed to show that they came from the ancient past, from a time long before mammals. But as far as anyone knew, there had never been such a thing as huge, plant-eating *reptiles*.

All the large herbivores that Mantell and his contemporaries knew— cows, horses, elephants, and so on—were mammals. A mega-crocodile munching its way through a field of ferns seemed as unlikely as a cow gulping down rabbits and foxes. Whose teeth had Mantell found?

Or whose teeth had Mary Ann Mantell found? As Gideon Mantell sometimes told the story (he set down several versions), it was his wife who had found the first of these mysterious teeth. She often accompanied Gideon on his medical rounds. One morning in 1821, or perhaps even earlier, Mary Ann waited outdoors by their carriage while Gideon was indoors with a patient.

Something on the ground caught her eye. The roadbed was made of rocks from a local quarry, broken into small pieces. Mary Ann Mantell spotted a small, dark object embedded inside one chunk.

A closer examination revealed curious features—ridges along the object's length and serrations along an edge. This was the tooth of an ancient, unknown animal, Gideon Mantell declared as soon as he saw it, and he soon found that the quarry it came from was the same one that his giant bones had come from.

(Historians now question this romantic tale. They believe the tooth likely came from a quarryman whom Mantell paid to keep his eyes open for anything out of the ordinary. It was indisputably the case, though, that Mary Ann Mantell spent endless hours—too many, she would eventually decide—on fossil expeditions with her husband. She had many finds to her credit, including several more teeth like this first one, and she became a skilled illustrator who prepared careful drawings of their best specimens.)

In 1822, Mantell traveled to London to show his strange teeth to the experts at the Geological Society. He carefully unwrapped his prizes and handed them over for examination. The teeth—brownish, with ridges on the sides, about an inch long—were not eye-catching, and the verdict came quickly. Mantell wrote disconsolately in his diary that his would-be trophies had been deemed of "no particular interest."

Perhaps they came from a fossil fish, William Buckland and his colleagues suggested, or perhaps Mantell had somehow managed to mix up recent relics from a rhinoceros or a hippopotamus with his ancient fossils.

In 1823, Mantell tried a different tack. Charles Lyell, a young geologist destined to win fame in years to come, planned to visit Paris to meet with Georges Cuvier. This was the scientific equivalent of a pilgrimage to Rome—Cuvier was the pope of paleontology—and Mantell prevailed on Lyell to show one of his fossil teeth and some select bones to the great man.

The new approach yielded the old result—Mantell's tooth came from a rhinoceros, Cuvier declared, and the small bones perhaps from a hippopotamus. Mantell sank into despair. Too miserable even to record the details in his diary, he summoned just enough strength to ask himself if there was any point in recording "mementos of wretchedness."

In time he regrouped. He began by taking a closer look into the geology of Sussex. His fossils, Mantell could soon demonstrate, had indeed come from rocks that had been laid down in ancient times. That made the case for rhinos or hippos hard to uphold.

He never managed to find a jaw—that would have settled matters at once, because mammals' jaws have different-shaped openings for the different-shaped teeth, and reptiles' jaws don't—but he had done the next best thing. Mantell had assembled a collection of teeth that showed different stages of wear and tear. Some were pristine, and some were worn down almost to nubs.

Paleontologists love teeth, which outlast bone. More important, they are telltale. Even a child can look in a dog's mouth, see the gleaming canines, and deduce, "Maybe Buddy could tackle a dinner that put up more of a fight than Yummy Kibble Bites." In somewhat the same way, paleontologists can often read a life's story in a tooth.

In the spring of 1824 Mantell sent Cuvier a package containing several of his still-unidentified fossil teeth. Cuvier pored over them, and then he did a remarkable thing—he changed his mind.

CHAPTER 15

Into the Temple of Immortality

On June 20, 1824, Mantell received a letter that marked one of the high points of his life. "These teeth are certainly unknown to me," Cuvier wrote. They did not come from a carnivore, and they looked as if they came from a reptile. "Might we not have here a new animal, an herbivorous reptile?"

Now, with this game-changing endorsement, all that remained was finding out just what sort of "new animal" Mantell had discovered. Clutching his precious fossil teeth, Mantell headed off to London and the Hunterian Museum of the Royal College of Surgeons, in the hope that something in its vast collection would throw light on his mystery.

The Hunterian was an immense and long-neglected natural history museum. It was named for its founder, John Hunter, a surgeon who had amassed a horde of more than ten thousand skeletons, fossils, teeth, bones, preserved animals, and anatomical bits and pieces of every sort, many of them horrifying. Hunter was reputedly the model for Mary Shelley's mad scientist, Victor Frankenstein. (Her novel had come out six years before, in 1818.) He was, if not quite mad, at least deeply eccentric.

His museum was a hodgepodge, with items arranged according to a scheme of Hunter's own devising. One cabinet featured a squid, slugs, a woodpecker, and a vulture, among others. Nearby was "a pile-up of

Two iguanodon teeth, still embedded in rock,
from Gideon Mantell's collection

genitals (rhino, sparrow, mole etc.)," one historian writes, as well as "jar after jar of floating calamity," where every sort of unfortunate or malformed creature bobbed in a bath of alcohol.

Mantell and the museum's curator, William Clift, combed through endless drawers of reptilian teeth and jaws. To no avail. Then, finally, Clift's assistant spoke up. He had recently prepared a skeleton from an iguana, a three-foot-long tropical lizard that had been floating in one of the Hunterian's countless alcohol-filled jars. Might he venture to suggest that the gentlemen look at the iguana?

They did, and they quickly saw that Mantell's large fossil teeth looked strikingly like the iguana's small teeth in every respect but size. The shapes matched, and even the ridges matched. Better yet, iguanas were herbivores! Now Mantell had a drawing to place at the top of his Most Wanted poster.

The fossil tooth was about twenty times bigger than the modern one, and Mantell's fossil bones seemed roughly twenty times bigger than the corresponding iguana bones. He had found a prehistoric beast, Mantell decided, that was essentially a lizard scaled up to gigantic proportions. He dubbed it iguanosaurus but soon rechristened it iguanodon (for "iguana tooth"), and this is the name still used today.

Mantell reckoned that iguanodon measured some sixty or seventy feet long, perhaps even one hundred! Even if that turned out to over-state matters a bit, he noted proudly, he had still found "one of the most gigantic reptiles of the ancient world, and a colossus in comparison to the pygmy alligators and crocodiles that now inhabit the globe."

In November of 1824, with Cuvier's letter clasped in his hand, Man-tell attended a meeting of the Geological Society. "What a pleasure it must then have been for him to confront his former critics," writes Man-tell's biographer Dennis Dean, still aggrieved on behalf of his hero nearly two centuries on. Especially, Dean adds, considering that "two years earlier they had sneered at his epoch-making fossils and, rather openly, at himself—the non-university son of an unlettered shoemaker."

Gideon Mantell

Now, finally, Mantell had something to write in his diary. "I shall ride on the back of my Iguanodon into the temple of immortality!" he exulted.

It made a happy interlude in a life beset by bad fortune, and the iguanodon did, almost at once, carry Mantell into the Royal Society. "It was with no small degree of pleasure," he confided to his journal, "that I placed my name in the Charter book, which contained that of Sir Isaac Newton and so many eminent characters."

Mantell's pride was well earned—he had overcome, Dennis Dean notes, "the virtually unanimous disapproval of his most esteemed colleagues"—but it is easy to misunderstand his achievement. It was not just that he had fashioned a dinosaur from a few discolored teeth and some fractured bones, though that remarkable feat was the ultimate in solving a jigsaw puzzle without looking at the picture.

The genuine coup was imagining such a beast in the first place. To *dig up* a dinosaur would have been a feat; to *dream up* a dinosaur was better still. Mantell's true achievement, the historian Ralph O'Connor argues, was keeping his intellectual bearings while the world tilted and tipped around him.

"To call Mantell 'the discoverer of the dinosaurs,' as if this conceptual category were sitting in the Tilgate quarry waiting to be 'discovered,'" O'Connor writes, "is to miss the literally unspeakable strangeness Mantell experienced as he grappled with the notion of a herbivorous lizard larger than the largest elephant."

And he would do it again. "I have made a grand discovery," Mantell wrote in his journal in 1832, and so he had. In the course of blasting apart rocks, quarrymen had spotted what one of them called "a great consarn [i.e., jumble] of bits and bones."

They contacted Mantell, who carted home the rubble from the explosion. He had no great hope—the quarrymen had loaded him up with a heap of fifty or so hunks of bone-containing rock—but he set to work cleaning the debris and trying to fit the pieces together.

In time he managed to assemble a stone slab roughly four feet by two feet that contained a dozen vertebrae, some ribs, bits of a skull, and other bony odds and ends. He set aside the fragments he could not find a home for; they filled a wheelbarrow.

Mantell soon convinced himself that he had found something new, though at first he thought he'd uncovered another megalosaurus or iguanodon. He had, in fact, discovered a new dinosaur.

This was Mantell's second dinosaur and the first armor-plated creature ever discovered. He named it hylaeosaurus, meaning "forest lizard."

The prehistoric world was beginning to fill up. Mary Anning's ichthyosaurs and plesiosaurs had roamed the seas. Pterosaurs patrolled the skies (the biggest *darkened* the skies, with wingspans that measured eight feet across). Now William Buckland and Gideon Mantell had sent the first land creatures rumbling into the picture.

Scientists and artists took a strange delight in describing not just a teeming prehuman world but a dog-eat-dog world, except that the dogs were the size of dragons. (Iguanodons were plant eaters, but that did nothing to keep them off the carnivores' menu.)

How did that dismaying picture fit with Paley's "happy world"?

This 1863 drawing, called *Earth Before the Flood*, by the French illustrator Édouard Riou, shows an iguanodon (at left) battling a megalosaur

CHAPTER 16

Framed for Bliss

For Buckland and Mantell, who were both skilled popularizers, this was tricky territory. They depicted the bygone world with awe and wonder, but even they had a hard time suppressing a shiver.

Proud as he was of his iguanodon, Mantell drew on a brand-new and frightening analogy. "Like Frankenstein," he wrote, "I was struck with astonishment at the enormous monster which my investigations had, as it were, called into existence."

The prehistoric world was marked by "perpetual warfare and incessant carnage," Buckland conceded, but he hurried to make an argument for a different point—contrary to appearance, this was cheery news.

Mantell focused less on carnage and more on strangeness. "No country on the face of the earth" looked like the prehistoric world, he wrote. "Nowhere is there an island or a continent inhabited by colossal reptiles only, or where reptiles usurp the place of the large mammals." Even the prehistoric graveyards, or what passed for graveyards, had an eerie look. "The bones of reptiles, of enormous size, are the only animal remains that occur in any considerable number."

The passage comes from a book on nature's "wonders," but the tone of horror and bafflement sounds straight out of Sherlock Holmes. "*Mr. Holmes, they were the footprints of a gigantic hound.*"

But Buckland and Mantell, who were friends, made a point of extracting a sunny moral from this dark story. Their reasoning was as straightforward as a proof in Euclid.

Prehistoric life was gory, they conceded, but that wasn't the heart of the matter. The essential point was that life was a good thing. And if there was life, then there had to be death as well, in order to give room for new plants and animals to have *their* chance at life. So the only question was, what kind of death was best?

That was straightforward, too. "The most easy death," Buckland wrote, was "the least expected." Out of benevolence, the Creator had fashioned carnivores' fangs and claws as fast, efficient killing tools. When death came suddenly and without a long buildup, there was no fear, no lingering pain, no torment from hunger or injury or age. (Buckland hurried to point out that the arguments in favor of a death without warning applied only to animals, not humans. In the wild, where there are no hospices or hospitals, the best fate for a sick or aged animal is a quick finish.)

Look at the world in the right way, and lions and tigers and megalosaurs were instruments of mercy. If death was inevitable, let it be quick! A few decades before, a French physician named Joseph-Ignace Guillotin had made an almost identical argument in explaining the merits of his own invention.

But why not a world made up entirely of plant eaters? Buckland polished off the question in short order. For one thing, a vegetarians-only world would deprive the carnivores of their chance at life.

Buckland went further, blithely explaining the benefits to the *herbivores* of living amidst a host of ravenous, sharp-toothed predators. Without carnivores to keep down their numbers, the herbivores would outbreed their food supply. "The whole class of herbivora would ever be so nearly on the verge of starvation, that multitudes would daily be consigned to lingering and painful death by famine."

God had devised a better way. Instead of dwindling away in pain and solitude, even the puniest herbivore could make one final gift to the world, by "contributing its enfeebled carcass to the support of its carnivorous benefactor." (Buckland had anticipated, by two centuries, a comic strip called *Li'l Abner* that featured creatures called shmoos, who could imagine no greater pleasure than to jump in a skillet to become someone's dinner.)

To hear Buckland and Mantell tell it, the whole system was a tribute to God's kind heart. The argument seems vulnerable in a dozen places, but the point is that it was not really an argument at all. God's goodness wasn't the *conclusion* you drew from looking closely at the world. It was the starting point, the never-to-be-questioned axiom that guided all your observations and theories. (This was a separate issue from good design—the deer's swiftness or the eagle's eyesight testified to God's skill as a designer, but it was the harmony of nature that testified to his benevolence.)

So Buckland and Mantell and their peers looked at living animals and at fossils and nodded in satisfaction. "The feeble and disabled are speedily relieved from suffering," Buckland noted approvingly, "and the world is at all times crowded with sentient and happy beings."

This was a straight lift from Paley, who had put the identical idea in virtually identical words. It's worth noting that this was a sincere argument and not a rhetorical tactic meant to deflect scripture-based protests.

Nor were Paley and his intellectual allies naïfs who saw the world as knee-deep in buttercups and birdsong. Paley himself knew physical suffering all too well. During the very years he was composing his happy hymn to nature, he suffered agonizing pain from an intestinal disease for which no treatment was known. He dubbed his affliction "the Scorpion"—the sketchy medical accounts of the day make it impossible to identify Paley's condition—and it devoured him from the inside out. After three years of torment, it killed him.

But Paley and his fellow philosophers seldom looked closely at their own biographies. Instead they focused on the natural world. There they saw a marvelous arrangement that worked on two levels at once. For a single creature, a swift and painless death hardly counted against a lifetime of pleasure. For the living world as a whole, a few were dealt out of the game at every round, but—and this was vastly more important—the game itself continued forever, in endless rounds of benign harmony.

"Thus the great drama of universal life is perpetually sustained," Buckland wrote, and "the same parts are ever sustained by another and another generation; renewing the face of the earth and the bosom of the deep with endless successions of life and happiness."

Mantell endorsed the same view, which he called a "sublime truth." This was a new view. Earlier thinkers had painted a far darker picture of the world. Nature was rife with cruelty, and death scythed its way along, unslackingly and insatiably. John Donne, to choose one example of many, took up the theme in elaborate set pieces.

> Th' earth's face is but thy table; there are set
> Plants, cattle, men, dishes for Death to eat.
> In a rude hunger now he millions draws
> Into his bloody, or plaguey, or starved jaws.

Even two centuries after Donne, many voices sounded the same grim theme. David Hume, the Scottish philosopher, wrote that "the whole earth . . . is cursed and polluted. A perpetual war is kindled amongst all living creatures." Erasmus Darwin (Charles's grandfather) called nature "one great slaughter-house, one universal scene of rapacity and injustice."

But, in the early 1800s, those messages were unwelcome and went largely unheard. If Buckland, Mantell, Paley, and their peers had questioned their own argument that nature was a happy place, they might have leapfrogged years ahead and focused on competition and struggle rather than on design and perfect order. They might have beaten Darwin by decades. They didn't.

They had come *so* close. With hindsight, we can watch as they pick up precisely the puzzle piece they needed, only to set it down in favor of a piece that, even from across the room, we know will never work.

Why did they miss it? Did they, almost literally, whistle past the grave-yard because they had ventured near a truth too frightening to bear?

Or was it simply that it was impossible to conjure up an idea whose time had almost, but not quite, come? (Even Erasmus Darwin quickly backed away from his "one great slaughter-house" judgment. On further reflection, he wrote three years after his first salvo, the world was "a well-balanced, harmonious system" and "warfare was the exception rather than the rule.")

John Maynard Keynes once remarked, in a different context, that "the difficulty lies, not in the new ideas, but in escaping from the old ones." Here the most important of those old ideas was not just that the world was happy nowadays, but that it always had been.

"Ere man was called into existence," Mantell wrote, "this planet was the object of the Almighty's care, and teeming with life and happiness."

To modern ears, this constant talk of happiness sounds forced and uneasy. The insistence on bringing dinosaurs and extinct rhinos and hippos into the happiness tent seems especially dubious, but our scientists had no doubts, or at least none they admitted.

A self-taught geologist (and amateur poet) named George Richardson put the "happy dinosaurs" case in lyrical form. Richardson was the curator of the museum that Mantell had created to house his fossil collection. He began his poem with a couplet in honor of dinosaurs and all other bygone creatures. (Richardson and his contemporaries called prehistoric beasts "saurians," from the Greek word for "lizards," or "primeval lizards," or even "antediluvian monsters," because they had lived before the biblical flood. The word *dinosaur* had not yet been coined.)

Saurians were relics of ancient worlds like coins or necklaces from Greece or Rome, Richardson wrote, but they were better than any such curios because they had been shaped by a divine hand. Ancient beasts were "Fragments vast of lost creations, / Relics of earth's first formations."

Then came several lines on the sheer unfamiliarity of "these giant forms tremendous / Creatures wondrous, wild, stupendous."

But the strangeness of these prehistoric creatures, Richardson emphasized, should not distract anyone from the real point. What was truly important was that these early creatures shared a crucial property with all the forms of life that would come along later: "Differing from a world like this, / Each and all were framed for bliss."

The moral of the story—that animals had lived in perfect contentment from Earth's earliest days—was so important that Richardson finished his poem with a rousing ode to joy. The final line spoke to creation as a whole and put special weight on the crucial word *all*.

> Form'd to share, without alloy,
> Each its element of joy,
> By that Power that rules to bless,
> All were made for happiness!

CHAPTER 17

"A Delicate Toast of Mice"

In the chorus proclaiming the news of a happy world, one voice stood out. It belonged to William Buckland. In an age that made a fetish of eccentricity, Buckland may have been the greatest eccentric of them all.

Buckland had a distinguished record—he had discovered megalosaurus; he was one of the leading geologists of his day; he was a professor at Oxford—but he was as far as could be from a staid academic. Even today one modern historian can scarcely bring himself to describe Buckland's antics.

His was "not behavior that at once commands respect," writes David Allen with dismay, and Buckland's fondness for "childish jests and puns" only made matters worse. He "lacked altogether the demeanor of the typical geologist: there was no hint of distant reverberations in his personality, no suggestion of ageless time, no ring of iron on stone; all people saw was a kind of learned clown."

Some people saw a sorcerer rather than a clown. One geologist—this was Roderick Murchison, the fox hunter turned scientist we met earlier—described his first peek at Buckland in his Oxford lair. "Having, by direction of the janitor, climbed up a narrow staircase," Murchison wrote, "I entered a long, corridor-like room, which was filled with rocks, shells, and bones in dire confusion, and in a sort of sanctuary at the end was my friend in his black gown, looking like a necromancer, sitting on one rickety chair covered with some fossils, and clearing out a fossil bone from the matrix."

Buckland's rooms were part attic, part menagerie. Cages full of snakes and frogs filled the halls. Fossils were stacked on every table and chair. In the dining room, candles had been arranged along an ichthyosaur's backbone.

A jackal wandered free. One visiting don, who made a point of sitting with his legs tucked beneath him to keep out of jackal range, "heard the animal munching up something under the sofa." He alerted Buckland. "'My poor guinea pigs!' he exclaimed, and sure enough, four of the five of them had perished."

Large buildings on the grounds outside were home to a fox, rabbits, ferrets, hawks, owls, a magpie, and a jackdaw, as well as dogs, cats, and chickens. A pet tortoise was so big that a small child could climb on his back and proceed on a (very slow) tour of the garden.

Buckland and his wife were perpetually gathering their children and heading off to hunt for fossils, or birds' nests, or moles, or to collect irises or lilies. (One son, Frank Buckland, grew up to become the David Attenborough of his day.) The family horse was so accustomed to these errands that it stopped automatically to wait at every garden and every quarry on their route.

Buckland's fondness for nearly all God's creatures, both living and dead, was genuine, but he wasn't a dotty Doctor Dolittle. He had come along at a time when geology was taking its first steps toward establishing a history of the world, and Buckland was Oxford's first professor of geology.

Like many of his scientific peers, he was both a scientist and a clergyman. (Buckland was an ordained minister whose father had been a minister, and he would eventually become dean of Westminster Abbey.) He took for granted that the roles did not conflict; the mission of science was to explain and explore the marvels of God's creation. Buckland was not earnest, to the dismay of many of his Oxford colleagues, but he *was* serious.

Many of his supposed sins were simply innovations (in Oxford in the early 1800s, the line between the two was blurry). He believed that geology was better studied outdoors than in a lecture hall, for instance, and he led his students on "geological rides," on horseback, to quarries or caves or cliffs where the rocks told tales to the informed eye. On other excursions, the class set out by train so that Buckland could "point out and explain the several different formations we cross."

Long-winded formality was the style of the day (preferably with a bit of Latin thrown in), but Buckland referred to his specialty as "undergroundology." He had a gift for oratory and a showman's flair. At one public lecture delivered outdoors, in the early 1820s, he spoke to an audience of thousands who had gathered in an enormous limestone cavern that had been fitted with artificial lights. Buckland explained the geology of the region—here were iron ore, and coal, and limestone—and then finished off with a patriotic flourish. Was all this wealth "mere accident"?

"On the contrary," Buckland roared, "it in fact expresses the most clear design of Providence to make the inhabitants of the British Isles, by means of this gift, the most powerful and the richest nation on the Earth."

This was in keeping with Paley's doctrine that the world was a living, breathing testament to divine benevolence. Buckland did not go quite so far as to say that God was an Englishman, but, after all, you could hardly blame him for having favorites.* Buckland ended his song in praise of geology and England on a high note, and the crowd responded with an exultant shout of "God save the queen!"

---

* Buckland spoke for an entire culture that not only shared his view but took it as beyond dispute. A generation after Buckland, a long essay in the *Christian Review* spelled out the argument in detail. Coal was a gift from God—"His first gift to the Anglo-Saxon race." He had bestowed this favor on "a race of men energetic and enterprising," distinguished "by their mental and moral culture and by their hold on the pure gospel of Jesus Christ." Anyone could see that God had put his thumb on the scale: "About nine-tenths of the coal of the world have been thrown by the Creator into the hands of the Anglo-Saxon race." Then, as if to underline his point, God had made an additional gift to his favorites. This second gift was not coal but *gold*. It had turned up "almost simultaneously" in two places, the United States and Australia. Could anyone fail to notice that both countries were English speaking? Here was further evidence of "a design on the part of God the Redeemer, a benevolent design towards these nations themselves, and through them, to the whole race of man."

*   *   *

Even when he was obliged to stay indoors, Buckland could make a geology lecture into a rousing affair. This was partly a matter of self-interest. At Oxford, subjects like Greek and Latin were mandatory but science classes were optional; students paid attendance fees, which supplemented the lecturer's salary. Buckland was by far Oxford's biggest draw.

His audience roared in delight as Buckland lumbered about the stage, imitating a huge, ungainly iguanodon. To show how pterodactyls flew, he gathered up the long tails of his coat and leapt into the air.

He was a showman, but he did not have to feign enthusiasm. Virtually anything could set him leaping for joy. The undigested contents of an ancient reptile's gut were more than enough inspiration. "When we see the body of an Ichthyosaurus still containing the food it had eaten just before its death, and its ribs still surrounding the remains of fishes that were swallowed ten thousand, or more than ten times ten thousand years ago," he wrote, "all these vast intervals seem annihilated, and we are almost brought into as immediate contact with events of immeasurably distant periods, as with the affairs of yesterday."

He was even more compelling in person. One ex-student retained a lifelong memory of a Buckland lecture. "He paced like a Franciscan preacher up and down behind a long show-case, up two steps, in a room in the old Clarendon. He had in his hand a huge hyena's skull. He suddenly dashed down the steps—rushed, skull in hand, at the first undergraduate on the front bench—and shouted, 'What rules the world?'

"The youth, terrified, threw himself against the next back seat, and uttered not a word. He rushed then on me, planting the hyena full in my face—'What rules the world?'

"'Haven't an idea,' I said.

"'The stomach, sir,' he cried (again mounting his rostrum), 'rules the world. The great ones eat the less, and the less the lesser still.'"

*   *   *

Buckland proposed to eat them all. On the long list of his eccentricities, Buckland's gastronomic habits claimed top rank. Meals in his household were a kind of ongoing experiment, with most of Noah's ark on the menu. Hedgehog was "good and tender," in one guest's judgment, but crocodile was "an utter failure." In his autobiography, written in old age, John Ruskin looked back wistfully to his undergraduate years at Oxford, when he had shared many a meal with the Bucklands. "I have always regretted a day of unlucky engagement on which I missed a delicate toast of mice."

Buckland himself was almost impossible to faze. The worst thing he had ever tasted, he once declared, was mole, which was "utterly horrible." Later on, though, he decided that the worst of all was the bluebottle fly, a bright blue relative of the housefly that shares all its cousin's vile habits.

It was not just the cuisine in the Buckland household that put his visitors to the test. One family friend recalled trying to keep his cool, at mealtimes when he was a boy, "while the guinea-pig under the table inquiringly nibbled at your infantine toes, the bear walked round your chair and rasped your hand with file-like tongue, the jackal's fiendish yell close by came through the open window, the monkey's hairy arm extended itself suddenly over your shoulder to annex your fruit and walnuts."

At one time or another, Buckland seems to have tasted nearly everything you might think of, and much else besides. His adventures in eating extended far past the dining room.

On one trip to Europe, he visited a cathedral where the tour included a stop to see a "martyr's blood." These were dark spots on the ground that renewed themselves perpetually, never drying up. Buckland dropped to his knees and licked the stain. "I can tell you what it is. It is bat's urine."

On another occasion, Buckland visited a nobleman who lived in great splendor near Oxford. Lord Harcourt showed Buckland the preserved heart of Louis XIV, which he kept in a silver casket. A Victorian writer named Augustus Hare picks up the tale: "Dr. Buckland, whilst looking at it, exclaimed, 'I have eaten many strange things, but have never eaten the heart of a king before,' and before any one could hinder him, he had gobbled it up, and the precious relic was lost forever."

*   *   *

The strange dinners and many of the ventures into natural history were family affairs. Buckland's wife, Mary, was his closest collaborator. She was an expert on fossils in her own right, an accomplished scientific illustrator, and a deeply knowledgeable student of the living world.

Mary and William met in a scene that could have come straight out of a rom-com. He was forty-one, she was twenty-eight, and the two of them happened to be traveling in the same coach. William Buckland was reading a thick, brand-new book by Georges Cuvier, the greatest anatomist of the age.

Buckland's fellow passenger was absorbed in a thick book of her own. Buckland was carrying a letter of introduction to a young woman who lived nearby, a Miss Morland, who was apparently deeply devoted to science. He ventured a word. *May I ask the name of the book that has so captivated you?* Could anyone fail to guess that the young woman would herself be Miss Morland? And that the book she was reading was the very tome by Cuvier that Buckland himself was reading?

The two were soon married. Well-to-do English couples in this era often took extended honeymoons in Europe. Trips might stretch out for months or even longer. ("Rather more than two years had glided away on the Continent," a biographer tells us, before Buckland's geologist friends Roderick and Charlotte Murchison turned their thoughts from celebrating their marriage toward returning home.)

William Buckland and Mary treated themselves to a ten-month honeymoon, heavy on geology (including a visit to Cuvier, in Paris). Mary kept a journal, but years later, one of her daughters remarked that "the scenery and associations of the spots visited are perhaps less carefully described than the character of the rocks."

The Bucklands raised nine children (four died young) and kept up their scientific collaboration throughout their marriage. On one occasion it occurred to William that a set of mysterious fossil footprints might perhaps have been made by an extinct tortoise. William woke Mary (it was two in the morning), who hurried into the kitchen and made a thick paste that she spread across the table.

William fetched their tortoise from the garden. Together they set him ambling across the tabletop. To their delight, they saw that the brand-new tracks and the ancient ones were a perfect match.

CHAPTER 18

Kirkdale Cave

Buckland's presentation of the megalosaur to the world, in 1824, was perhaps the highlight of his professional life. But his word carried such weight because a triumph a few years earlier had rocketed him to fame.

The story began with a cache of mysterious bones—hyena bones prominent among them—that had been found deep in a Yorkshire cave in 1821.

This was strange from the get-go, since no one had ever seen hyenas in England. Buckland hurried to the site and crawled his way about a hundred yards through a small, twisting passageway that squeezed down, at its narrowest, to about two feet by two feet.

Then the passage opened up. Buckland struggled to see in the gloom, with a lantern as his only light. Suddenly, the mud and muck revealed their secrets. "The bottom of the cave . . . was strewed all over," Buckland wrote, "from one end to the other, with hundreds of teeth and bones."

The quarrymen who had discovered the cave had assumed the bones came from cattle that had died a few years before. That was a natural guess, but it proved profoundly wrong. Buckland eventually identified bones and teeth from some two dozen species, including hyenas, elephants, rhinoceroses, hippopotamuses, deer, foxes, birds, and bears.

Many of the bones were half-buried in mud. They poked upward, Buckland remarked (with food never far from his thoughts), like pigeon legs through a pie crust.

Little of this made sense. Why were so many bones heaped together in a common grave? Why so many species? These were, after all, animals that would never have voluntarily shared a refuge. And how had warm-weather creatures like elephants and hippos made their way thousands of miles from their sunny homes to end up in one of England's coldest, draftiest corners?

Buckland's first thought was that he had found proof of the truth of the Bible story of Noah and the flood. For a great many thinkers of this era, not just Buckland, the flood was the go-to explanation for many of the world's strange features. When travelers found fossilized seashells high atop mountains or when skeletons from elephant-like mammoths turned up in places where elephants did not belong, like Siberia, no one was much puzzled: it was the flood that had done it.

The idea was straightforward, and scientists had happily invoked it for centuries. As Noah bobbed along in the ark, they explained, flood-waters had risen higher and higher across the globe. Colossal waves had swept up animals from all over and flung them around willy-nilly. Eventually the waters had subsided and left shells and skeletons behind.

This was a tidy theory and an appealing one, too, since it so neatly reconciled the newest scientific findings and the oldest religious teachings. Here was clear proof of the long-established doctrine that the word of God and the works of God told the same story.

In the case of Kirkdale Cave, as it was known, an inconceivably powerful surge of water had flung a miscellany of animals headlong into a deeply hidden cavern. This was Buckland's first notion as he turned his lantern this way and that in the darkness.

But he soon came to question his own suggestion. How, Buckland asked himself, could enormous animals like elephants have ended up inside a cave whose only entrance was a tight, twisting passageway barely big enough for a human being to wriggle through?

His doubts grew when he examined the bones more carefully. "Scarcely a single bone has escaped fracture," he noted, and the bones were not just broken but covered with tooth marks.

The flood theory could explain the fractures—mighty waves could have smashed animals against one another and then shot them out against hard, sharp rocks—but who had done the chewing? These bones looked more like the remains of a scavenger's feast than the skeletons of drowning victims.

Buckland observed something else that was odd. A great many of the teeth and bones, he had noticed at once, came from hyenas. Later, after he'd had time for a careful count, he would guess that the cave contained remnants of two hundred to three hundred hyenas. None of the other species turned up in anything like those numbers.

Buckland proposed a new theory: The animals in Kirkdale had *not* been thrown into the cave by the biblical flood. Nor had they been transported halfway across the globe. They had lived in England long *before* the flood (when, presumably, the climate had been different). That fit with other clues that scientists were turning up at about the same time, like Mantell's fossilized palm trees, all showing that Europe had once basked in tropical warmth.

Kirkdale Cave was a hyena den, Buckland now proposed, and not a biblical relic. Over the course of many years, he explained, hyenas had killed or scavenged countless animals, dragged bits of their carcasses into their den, and scarfed down their remains.

But there was a problem. It was Buckland himself who was the most prominent spokesman for the flood theory, the doctrine that he had now rejected.

This was a predicament, and a predicament on several levels at once. It was trouble for Buckland personally, for one thing, because the Kirkdale discovery was headline news. No one would miss this story, and no one would miss Buckland's U-turn.

"You have just made a discovery, a fantastic discovery, in a cave in Yorkshire," as the historian Mott Greene summarized events. "This discovery is going to make you the leading scientist of your generation. It will make students and faculty jostle for seats in your lecture hall. It will make you a best-selling author and the subject of not only newspaper and magazine profiles but long articles in learned journals. It will lead to prizes, offices, money, honors, and fame."

The catch was that Buckland had spelled out the old view just two years before, in 1819, in a book entitled *Vindiciae Geologicae* [Geology Vindicated], *or The Connexion of Geology with Religion Explained*. The book was a reprint of Buckland's first lecture at Oxford, to celebrate his appointment as Oxford's first professor of geology.

Geology was "vindicated," he had written then, because the biblical account of creation was "in perfect harmony with the discoveries of modern science."

And now it wasn't.

This was not a small matter. Science in this era was viewed with deep suspicion, on the grounds that it threatened to undermine faith in religion. Buckland saw his mission as demonstrating that the distrust of science was misplaced.

At Oxford, Buckland's academic home, that distrust was widespread. Oxford had only made the decision to include geology in the curriculum with fear and hesitation. In the early 1800s, science and religion were merged in a way that scarcely exists today, and religion was the dominant partner in that union.

Oxford was almost as much a theological seminary as a university, one modern historian observes. "Science teaching was not intended to institute a modern, professional education" at either Oxford or Cambridge, a second historian agrees, "but rather to educate Christian gentlemen." It worked. "Half of the students became clergymen, and most of the college fellows were in holy orders."

Buckland's impeccable religious credentials made him a perfect choice to educate those young gentlemen. He believed with all his heart that geology testified to God's grandeur. The creator was the "Omnipotent Architect," Buckland had written in *Vindiciae Geologicae*, and every hill, valley, and cliff offered rock-solid evidence that He had arranged matters with magnificent care.

Buckland was, simultaneously, a devout Christian and a careful and conscientious scientist who believed in following the evidence wherever it led. He'd never had reason to question either of his core beliefs. On the contrary, he'd roared to fame, basking in applause like the acrobat in an old-fashioned circus who came racing across the ring standing atop two horses galloping side by side, one foot on each.

For Buckland, one horse was science, the other religion. Suddenly they seemed to be veering apart.

Now what?

Buckland's search for a way out began not with a dive into theology but with a detour into zoo-keeping. Buckland was a good observer with a knack for devising ways to test his theories—anyone who would lap up bat urine plainly had an experimental bent—and he got hold of a young hyena that had recently been captured in Africa and brought to England. "Billy" joined the Buckland menagerie for a time.

Buckland watched with delight as his new pet cracked open bones (from an ox), extracted the marrow, and then devoured the entire bone except for the knobby bit at the end. "Billy has performed admirably on shins of beef," Buckland exulted in a letter to a friend, "leaving precisely those parts which are left at Kirkdale. . . . It is impossible to say which bone had been cracked by Billy and which by the hyenas of Kirkdale!"

(Buckland had planned to finish up his experiment by killing the hyena so that he could compare its skull with skulls from Kirkdale, but he couldn't bring himself to do it. After his stint as a pet, Billy retired to the Surrey Zoological Gardens, in London, where he reigned as a celebrity and lived to an old age.)

More good news came from India. Modern-day hyenas there, one scientist reported, were so similar in habits and anatomy to Buckland's prehistoric hyenas that anyone would think "they had attended regularly three courses of his lectures."

For Buckland's contemporaries, his picture of Kirkdale Cave was not just a neat bit of historical detective work but something far more startling. Other people had pieced together ancient bones and reconstructed isolated creatures from the vanished past. Buckland had brought a whole *scene* to life, with a caveful of prehistoric creatures warning away their rivals and tearing off chunks of flesh from their victims.

The truly startling bit was that this was a scene that no human eyes had ever beheld. A geologist friend of Buckland's made a drawing of him poking his way into Kirkdale Cave, holding a candle, and gazing around in wonder at living, snarling hyenas.

This was time travel. In 1822, decades before Jules Verne and H. G. Wells dreamed up the first time machines, it was a new and head-spinning notion. Every kind of story *except* time travel had been known since humans had first huddled around fires in caves and listened spellbound as bards spun sagas. Love stories, adventure stories, and murder mysteries were all old news. But until life suddenly accelerated and transformed itself in the 1800s, no one had ever wondered about what life would have looked like in a different era. Why ask, when everyone knew? It would look like it had *always* looked. In the past and in the future, men and women would scrape out a living as best they could. They would dig in the dirt to grub out a rough meal as their parents had and their parents before them, and as their children would and their children's children after them.

The cave drawing was a gift between friends, not a work meant for the public, but it was one of the earliest attempts to imagine prehistoric animals in action. (The artist was William Conybeare, the geologist who had presented Mary Anning's plesiosaur to the world.)

More polished illustrations on the same theme would soon follow. We have already encountered the best-known, a watercolor of Mary Anning's "monsters" fighting and flying. But it was the decision to inject Buckland into the scene, and at center stage, that grabbed early viewers by the throat and made their eyes bug out.

In 1822, the Royal Society awarded Buckland its Copley Medal, akin to a Nobel Prize today, for his Kirkdale work. It had never before gone to a geologist. Buckland would keep his place among Europe's scientific elite for the next two decades.

He would go on to open up vast eras of prehistory. He would convince the world that Earth was ancient and that it had once been home to animals far stranger than the elephants and hippos in Kirkdale Cave.

He would help to show that geological mysteries—immense boulders sitting on their own in the middle of nowhere, for instance, and deep, parallel gouges cut into rock surfaces—were *not* evidence of the flood.

Buckland publicly rejected his old view in favor of a new theory that sounded equally preposterous. In the 1830s and '40s, when a Swiss scientist named Louis Agassiz proposed that the world had been subject to periodic deep freezes—this was the first mention of ice ages—Buckland became one of the earliest converts to the new doctrine.

The idea was startling, but the evidence was undeniable. Vast regions of the world had once been covered with sheets of ice a mile thick, Buckland argued, and it was those glaciers, not a flood, that had reshaped the world.

In his prime, Buckland was celebrated and greatly admired. He had found a way to deliver a message that was both entertaining and inspiring, and his professional peers and the public joined in showering him with honors.

But his story had a grim ending. In his midsixties, Buckland began acting in strange and worrying ways. (In hindsight the episode when he gulped down Louis XIV's heart may have been a sign that something had gone badly wrong.) Though he had always had a gift for friendship, Buckland retreated into silence and took to slapping himself on the head and scratching himself fiercely and compulsively. The odd behavior persisted and worsened.

Trapped by strokes or disease or senility, Buckland seemed beyond reach. Doctors never managed to come up with a diagnosis. The guessing game continues to this day, and the paleontologist and historian Tim Flannery has even suggested that perhaps Buckland's adventures in eating caught up with him and he fell victim to mad cow disease.

Mary Buckland sought help from everyone she could think of, but no one could offer much beyond prayer. In the end, she committed William to a lunatic asylum near London, where he spent his final days, in a friend's troubled words, "amongst outrageous madmen."

CHAPTER 19

In Nature's Cathedral

Looking back, we might find it surprising that Buckland rose to fame. Today the notion of a celebrity geologist is hard to take in. At the time it seemed perfectly natural.

Not every culture would have turned geology lectures into can't-miss entertainment or written poems in honor of dinosaurs. Not every culture would have sent Mary Anning and a host of rivals scanning the ground in search of fossils. But the Victorians did. In England in Mary Anning's day, and for decades afterward, people at every rung on the social ladder were obsessed with natural history to a degree we can scarcely fathom. Fossils were only part of the picture.

The whole country seemed possessed, in one historian's words, by "an overwhelming drive to collect, witness, and catalog nature." Victorians (and their Regency predecessors) were crazy about ferns and seashells and insects and birds' eggs and mushrooms.

Every field and meadow was dotted with enthusiasts in pursuit of butterflies or caterpillars or cocoons, with a net in their hands or jars tucked under an arm. Every tide pool was surrounded by damp collectors on the watch for starfish or anemones.

Inanimate wonders like geodes and crystals and fossils won hordes of admirers, and nearly everything that crept or crawled or grew in the soil had its devotees. Beetles, with their striking colors and endless variety, were especially popular. Beetle-mania struck nationwide; at Cambridge, beetle collecting grew "just as competitive as cricket or rowing."

No facet of the natural world was too mundane to induce rapture. "My heart leaps up when I behold / A rainbow in the sky," Wordsworth exulted in 1807, but it didn't take rainbows to get his pulse racing. A few years earlier, he'd written a poem about showing his beloved a glow-worm—a kind of cousin to a firefly.

> The whole next day, I hoped, and hoped with fear;
> At night the glow-worm shone beneath the tree;
> I led my Lucy to the spot, "Look here,"
> Oh! joy it was for her, and joy for me!

For Wordsworth, and for the English generally, the natural world was not simply beautiful. More important, it was uplifting. Wordsworth found in nature "a presence that disturbs me with the joy / Of elevated thoughts."

Aldous Huxley would write later that Wordsworth (and his countrymen) would never have arrived at this cheery view had they lived in a harsher part of the world. The English landscape was benign; it was easy to indulge in lofty thoughts when you had no experience of avalanches or volcanoes or earthquakes, or lions and tigers and pythons.

In England, Huxley wrote, "a walk in the country is the equivalent of going to church." The tropics were different. "The jungle is marvelous, fantastic, beautiful; but it is also terrifying, it is also profoundly sinister."

For centuries the English had taken for granted a tamed, subdued nature. ("It droppeth as the gentle rain from heaven," wrote Shakespeare, who had never witnessed a monsoon.) The difference between the Lake District and the Amazon was akin to the difference between a house cat dozing in an armchair and a jaguar leaping on its victim.

For the English, in the 1800s, nature beckoned and reverence came easily. Guidebooks of all sorts sold out as quickly as they could be printed. A book called *Common Objects of the Country* sold a hundred thousand copies in a week. Those were John Grisham numbers (in a much smaller country) for a book about frogs and bats and toads and weasels. "Every young girl," the historian Lynn Barber writes, "had at least twenty names of ferns, mosses, fungi, and beetles at her fingertips."

No one was too young for an immersion in natural history. An immensely popular author named Margaret Gatty specialized in books for children. One of her best-known, *Parables from Nature*, went through one hundred editions. She found inspiration under every leaf. "Nature abounded in 'wonderful adumbrations of divine truths,'" Gatty wrote, such as the transformation of a lowly caterpillar, enclosed in a dark cocoon, into a glorious butterfly, soaring into the light.

The fever swept up all sorts of strange bedfellows. "Almost everyone— scientist, novelist, artist, poet, musician, and, of course, people of humble birth and no profession—seemed primed to respond conspicuously to the newly revealed wonders of the Earth," the historian Michael Shortland marveled.

Shortland had in mind wonders literally of the earth, like fossils, rather than natural wonders generally. That makes sense because, curiously, it was geology, perhaps more than any other subject, that fascinated the public in the nineteenth century. No one has ever quite managed to explain why, though there was a vague but widespread belief that geologists could see deeper than others could.

Geologists themselves never made that claim in so many words, but again and again they invoked the image of the earth as a book that they had learned to read. The earth had a history, just as nations did. This history was inscribed in stone in a difficult but not impossible language.

In precisely this era, scholars had scored a colossal coup—Jean-François Champollion had deciphered the hieroglyphs on the Rosetta stone—and now thinkers applied that deciphering analogy everywhere. Fossils and prehistoric lizards, one poet wrote, were "Wondrous shapes, and tales terrific / Told in Nature's hieroglyphic."

Fossils had special pizzazz because they helped give a shivery sense of the vastness of time. A fossil that you could hold in your hand made the mystery of endless time almost tangible, in the same way that a hunk of meteorite that you could touch and heft might convey something of the depths of space.

Geology was sexy, though no Victorian would ever have used that word. (It's not true that Victorians covered up piano legs because bare legs were too risqué, but they did have a knack for spotting dangers lurking in the most innocuous settings. One well-known scientist had to tiptoe around the word *mammal* in a lecture, for instance, because it veered too near *mammary*. Mammals were distinguished from other animals, he explained, "not only by having living young, but by nourishing their young in a peculiar way.")

The public, not just scientific professionals, gulped down geological books. Tennyson devoured Charles Lyell's *Principles of Geology* and then thought nothing of larding a poem (about a princess's adventures) with allusions to "shale and hornblende, rag and trap and tuff, / Amygdaloid and trachyte."

The notion that a poet might be a science buff didn't strike anyone as odd or even especially noteworthy. While in his twenties, Tennyson had set himself a program of study. Tuesday mornings were for chemistry; Wednesday mornings, botany; Thursday mornings, electricity; Friday mornings, animal physiology; Saturday mornings, mechanics.

The message of Lyell's text was that geology shaped the world ever so gradually, as wind and water worked their slow-motion changes. Tennyson happily incorporated such notions into immensely popular works of poetry:

> The hills are shadows, and they flow
> From form to form, and nothing stands;
> They melt like mist, the solid lands,
> Like clouds they shape themselves and go.

Tennyson had additional geological lessons to impart. Not only had hills melted, but land and sea had changed places, like partners in an eternal dance:

There rolls the deep where grew the tree.
O earth, what changes hast thou seen!
There where the long street roars, hath been
The stillness of the central sea.

Painters as well as poets fell under science's spell. In the early 1800s, one historian writes, "scientists, poets, and artists felt that they were traveling forward together . . . towards new horizons, and there was no atmosphere of compartmentalism or of barriers between cultures, the world of science and the arts intermingling as a matter of course."

What was more unexpected was that the fervor for science embraced both the working-class and the well-to-do. The pursuit of science was "essentially egalitarian," in the judgment of the historian Barbara Gates, who does not quite succeed in subduing the note of surprise in her voice. "No one was barred from its pursuit. Local clubs were formed by workingmen; women interpreted science for women and children; and a huge industry of cheap publications touting natural history arose."

Lectures on electricity or chemistry for the general public drew standing-room crowds. Humphry Davy, a renowned chemist and a captivating speaker, won rounds of applause with showy demonstrations that featured explosions and bursts of fire. Aristocrats competed for the chance to take the stage and serve as Davy's assistant. *Would you be so good as to hold this for a moment?*

"Tickets for the lectures were almost impossible to secure," one historian writes, "carriages jammed Albemarle Street on lecture evenings, and Davy became the lion of the day and London's most sought-after intellectual dinner guest." Davy was a small man with a soft voice, but he was handsome and had dark, intense eyes. Women made up half the audience and sent Davy mash notes and presents.

Gideon Mantell, the fossil finder who had discovered the iguanodon, was another lecturer who drew large, enthusiastic crowds. Mantell's listeners followed his excursions into natural history wherever they led. "Ladies of rank and fashion were seen handing round glasses containing dissections of the eyes of sheep, oxen & etc," one observer remarked, "and examining them with as much interest as the contents of caskets of jewels often excite."

This was the backdrop for the dinosaur discoveries. Here was a culture that swooned at a firefly. No wonder they were bowled over by word of a ten-ton lizard.

CHAPTER 20

"Quite in Love with Seaweeds"

Whole nations have been swept up in fads, like Holland's tulip craze, in the 1600s, when a single bulb could cost more than a house. But the Victorian nature frenzy stands on its own, for several reasons.

Its duration set it apart, for starters. Fossils, to take one example, kept their allure throughout the nineteenth century. Mary Anning sold her first fossil in 1810. Sixty years later, in a letter to a friend, the novelist Charlotte Yonge wrote happily that she had to sign off in a hurry "in consequence of an invitation to go out and hunt fossils."

The breadth of the nature craze was special, too, as we have seen. Most important of all was its motivation. In a devout era, it was almost mandatory to find some way to insert religion into the story.

The Victorians' task was to reconcile two scenes that seemed not to fit together at all. One was the picture, just coming into focus, of a world populated with prehistoric lizards lumbering through swamps and munching on ferns. The other was the familiar world where sheep and horses and humans wandered across green fields. Plainly God had created both vistas, but why had he gone to such trouble? What was the divine plan?

While that perplexing question lurked unresolved, people focused their attention on the here and now. Large theological questions might be beyond the common ken, but anyone could pay homage, in a small way, to the wonders of God's creation.

As a result, the most mundane pursuits took on a kind of celestial glow. To partake in nature—even in as minor a way as to clamber over a hill or to fish in a stream—was to honor God. Here was a temple where everyone, male or female, educated or unlettered, prosperous or poor, could join in worship.

You didn't even need to venture outdoors. Victorian decor was big on clutter, and no self-respecting home was complete without a nature-related display. Ferns, orchids, and aspidistras were almost mandatory. Ornate cages with squawking parrots or cooing lovebirds were a must.

A brand-new sort of display case proved the most exciting and eye-catching item of all. Glass had always been too expensive to use in "frivolous" ways. Then, in 1845, a long-standing glass tax was repealed, and suddenly aquariums (and greenhouses) were all the rage. Large, crowded aquariums quickly became a status symbol.

Unfortunately, for reasons no one could fathom at first, fish tended to die soon after they were released into their tank. The credit for discovering that putting aquatic plants in the water made all the difference goes to a woman named Anna Thynne. In 1846 she'd spent a seaside holiday at Torquay, on England's southern coast. She collected some corals in a jar of seawater and brought them back to London (where her husband was the dean of Westminster Abbey).

Thynne did not come to her insight right away, though people had long suspected that fish in bowls needed their water kept "fresh" in some way. With no seawater near at hand, Thynne's first idea was to have a servant keep the water aerated "by pouring it backwards and forwards before an open window, for half or three-quarters of an hour" every day.

"This was doubtless a fatiguing operation," Thynne conceded, "but I had a little handmaid, who, besides being rather anxious to oblige me, thought it rather an amusement."

A year later, presumably to the relief of her maid, Thynne found that adding kelp to the tank ensured long, healthy lives for her fish.

*   *   *

Not only aquariums, but also seashells and everything else connected with the sea had huge appeal. Like a great many others, the novelist George Eliot proclaimed herself "quite in love with seaweeds."

Eliot and her partner, George Henry Lewes, liked to spend holidays geologizing at promising locales like the Isle of Wight, tapping rocks with their hammers in search of souvenirs. But they found seaweed as alluring as rocks.

So did countless others, including Queen Victoria herself. The craze spanned decades. As far back as the 1790s, Queen Charlotte, the wife of George III, had dazzled the guests at an elegant ball with a dress in a seaweed pattern by one of the era's most exclusive designers.

Seaweed admirers who did not have a fabric designer on hand gathered seaweed at the beach and carefully pressed their prize finds into albums. (Along with her children's books, Margaret Gatty wrote a two-volume opus called *British Sea-weeds*, with tips on collecting and hundreds of lovingly drawn color illustrations.)

As a girl, Victoria made a seaweed album that she gave as a gift to the queen of Portugal. Ordinary collectors donated their albums to auctions to raise money for good causes, as today we might donate a quilt or a pie.

This drawing of collectors at the seashore appeared in
the satirical magazine *Punch* in 1858. It's by John Leech,
one of the era's best-known illustrators.

Seashells drew perhaps even more admirers than seaweed. Shells were beautiful and eye-catching, especially ones from far-off lands, and their uses were endless—for display on shelves, as ornaments in elaborate mini-grottoes, for fashioning into brooches and other bits of jewelry. Dealers competed fiercely, and sailors newly arrived in port were accosted by buyers eager for new stock.

One entrepreneur's story offers some idea of just how popular seashells were. In the fall of 1833 a London couple, Marcus and Abbie Samuel, began selling "small Shells for Ladies' Work." Soon they moved on to selling "trinket boxes" to tourists at seaside resorts.

This was not a glamorous enterprise. The Samuels' shop was in London's cramped and crowded East End, where pawnbrokers and rag shops dotted the streets and large families crammed themselves into small, squalid flats. But the Samuels had found a niche. "The little boxes sold so well," the historian Cynthia Barnett writes, "that the Samuels soon added shell sewing boxes, shell needle cases, shell portholes, shell frames, and other varnished mementoes."

Soon there were forty women at work, churning out shell tchotchkes. The Samuels established far-flung connections, especially in China and Japan, for buying ever more seashells. A new generation took over the family trinket shop and renamed it. Now they were the Shell Transport and Trading Company.

The new name hinted at new ambitions. The business still exists. Now it is known familiarly as Shell Oil.

William Paley Stubs His Toe

So natural history had an appeal that stretched both wide and deep. But what makes our story so strange is that the dominant message of the age—*All is well, nature is benign, life is a lullaby*—required that you cover your ears and block out the new teachings of science: *Life is harsh, losers outnumber winners, every day reveals new evidence that life has been fierce and bloody from the beginning.*

The trick was finding a way to fend off the grim news while still paying homage to nature's wonders. The gory dinosaur news made the challenge all the more pressing. How could you do it?

One approach was to focus on beauty or curious facts rather than on deep questions. No one asked why there were so many species of animals, for instance, or how it might benefit a peacock to grow so magnificent a tail that it could scarcely fly.

Instead naturalists examined in great detail such wonders as a mole's agility. A mole could race through the tunnels in its burrows at high speed, one writer marveled, and, if need be, it could scootch its way backward nearly as quickly.

Nor did anyone pose awkward questions about why a benevolent God had made a world where ravens pecked out the eyes of helpless lambs or where more baroque forms of cruelty prevailed. Why, for example, had God ordained a system where wasps paralyze caterpillars and then lay eggs in the still-living bodies, so that the larvae can enjoy an endless bounty of fresh meat?

Questions like that would have spoiled the mood, as if someone at a banquet had stood up and denounced the working conditions in the restaurant kitchen. But if anyone *was* inclined to ask such pesky questions, William Paley had shown how such challenges could be fended off.

Later writers endlessly echoed Paley. His best-known argument stemmed from what seemed like an irrefutable analogy. "Suppose I had found a watch upon the ground," Paley began his opus *Natural Theology*. It would be natural to wonder how it had come to be there. Would anyone propose that it might have lain there forever, like a stone in a field?

On the contrary, Paley argued, as soon as you examined the watch, you would see how intricately engineered it was, with a myriad of tiny screws and springs and precisely cut cogs and gears, and all of them meshing perfectly. You would recognize at once that someone had *designed* this watch.

Who could disagree? But then Paley pounced. "Every indication of contrivance, every manifestation of design, which existed in the watch," he declared, "exists in the works of nature." Even more was true—the works of nature far *surpassed* even the most elegant of watches.

This was an argument with a long history. Many centuries before the first watch, the Roman philosopher Cicero had looked at sundials and anticipated Paley. "When you see a sundial or a water-clock," Cicero wrote, about a century before the birth of Christ, "you see that it tells the time by design and not by chance."

If someone had to design even something as barebones as a sundial, Cicero asked, how could anyone suggest that the universe itself, with its countless complex features, had arisen by chance?

Cicero lived in a world with only the simplest mechanical devices. In Paley's era, when complex new machines appeared almost daily, the analogy seemed even more compelling.

The argument for design was simple and seemed irrefutable. Paley was, moreover, a clear and lively writer, and his case seemed all the stronger because he presented those facts that best supported him and skated by those that might have caused trouble. This was the literary equivalent of a realtor's touring a client through a new house and making a fuss over the marble counter in the kitchen while hurrying by the dark, cramped bedrooms.

Paley began a chapter on plants, for instance, by noting that it was harder to see signs of design in plants than in animals, and so he would speed on to more promising territory. "It is unnecessary to dwell upon a weaker argument," he remarked, "where a stronger one is at hand."

That is not how science is meant to work. The hardest cases—*How does evolution explain the eye? Why is extinction so common?*—are precisely the ones that present the most important tests of a theory.

But the argument for design was easy to grasp and hard to rebut, even without Paley's storytelling sleight of hand. Earlier writers had done their best to shoot it down, but to little avail. Perhaps the most eminent skeptic was David Hume, the brilliant and slyly malicious Scottish philosopher.

The world was rife with pain and evil, Hume noted, which meant there was no good reason to presume that God was all-wise and all-powerful. Perhaps, he suggested, the world had been created by "some infant deity, who afterwards abandoned it, ashamed of his lame performance." Or perhaps a second-rate deity had been in charge. Or an over-the-hill deity whose shabby results were a product of "old age and dotage."

Few thinkers paid much attention, and ordinary men and women were even less perturbed. The argument for design was one that anyone could absorb and embrace. Best of all, the conclusion could be set out in a few reassuring words: God had designed all the countless inhabitants of the world.

From that starting point, it followed that any challenges to the natural order were misguided. We might fail to understand some feature of the world, but that was *our* failing. A perfect designer did not do imperfect work.

That was important, because it meant that rambles in search of butterflies or starfish were more than casual outings. England in the nineteenth century was a serious-minded, moralistic culture. A stroll without a purpose might have appeared dangerously self-indulgent.

Margaret Gatty, the great seaweed connoisseur, had come near acknowledging the simple pleasure of a day outdoors. In an era when women, especially, were constrained at every turn, her delight at the prospect of a seaside excursion leaps off the page. "To walk where you are walking, makes you feel free, bold, joyous, monarch of all you survey," she exulted, "untrammeled, at ease, at home!"

Safer, though, to couple that hint at rebellion with some quiet contemplation of God's handiwork. Fresh air was a fine thing, but an excursion to pay homage to God was far better.

Earlier generations had studied nature out of "blind curiosity," the naturalist E. P. Thompson scolded in 1845. Now that the focus had shifted to religion, prospects had improved. "The general tendency of the study is to lead us from the admiration of the works to the contemplation of their Author," Thompson wrote approvingly, "to teach us to look through nature up to nature's God."

This merging of God and nature had the further benefit of providing a one-size-fits-all philosophy that could draw in nearly everyone. The devout could hail God's handiwork. Those less inclined to talk of miracles and everlasting life could focus instead on the cycles of the seasons and similar signs of a smooth-running cosmos. Even religious skeptics, few though they were, could scarcely object to a credo that spoke more of eagles and elm trees than of doctrinal disputes.

But putting religion and science into one basket was risky, though no one at the time seemed to see the danger. Lynn Barber, a historian and the great authority on Victorian attitudes toward nature, spelled it out. "The Victorians saw nothing glorious in the pursuit of knowledge for its own sake," she wrote in a groundbreaking book called *The Heyday of Natural History*. "It was *only* religion, in the shape of natural theology, that made the study of natural history worthwhile."

Safeguarded by that stamp of approval, the study of nature took off. "If there had been any hint, at that stage, that it could lead to irreligious or anti-religious views," Barber writes, "nobody would have dreamed of taking it up."

Instead, both professional scientists and eager amateurs spent their lives in the happy belief that they were building a cathedral, never knowing that in fact they were erecting a tomb that would encase all that they held most dear.

Here Be Dragons (and Giants and Cyclopses)

Enormous bones had turned up many times through the ages, long before Mary Anning or William Buckland or Gideon Mantell ever grabbed their geological hammers and set to work. People had dug wells and farmed fields since ancient times, which meant they had occasionally happened on strange finds.

Rivers had carved channels across vast landscapes and offered peeks at what had long been buried. In England, for hundreds of years, miners near Oxford had ventured deep underground in search of limestone for slate roofs. They had unearthed countless bones as they dug.

So in theory someone might have shouted "Dinosaur!" many centuries before the 1800s. But that's unlikely, because discovering is not merely finding something; discovering is finding and understanding that you've found something. A dog could chase a duck into the Fountain of Youth, but we would hardly hail its "discovery."

And even if some ancient thinker had managed to conceive of a world ruled by vanished beasts, it seems unlikely that she would have won any followers. Not every era is ripe for every idea. "A given genius may come either too early or too late," William James wrote in 1897. "Peter the Hermit would now be sent to an insane asylum." What was true of an orator at the time of the Crusades applied equally well to a Greek warrior from the era of the *Iliad*. "An Ajax gets no fame in the day of telescopic-sighted rifles."

Most of the early finds were not dinosaur bones, we now recognize. Many came from mammoths and mastodons, which are now extinct but lived as recently as ten thousand years ago, or from prehistoric hippos and giraffes. Others were relics of imposing but rarely seen creatures that still survive, like narwhals and whales.

The dinosaur bones dated from tens of millions of years ago, or hundreds of millions. The mammoths lived so recently that they show up, beautifully depicted, in cave paintings. The bones from these various finds had only two things in common—they were big, and they were mysterious.

When we think about dinosaurs today, we unconsciously draw on two centuries' worth of work by scientists trying to put flesh on those intriguing bones. But nineteenth-century scientists who made their own tries at reconstructing past lives had no such models to guide their thinking.

Often they had only a few bones to work with and only the vaguest ideas about what a plausible beast might have looked like. Imagination ran as free as in a Rorschach test.

"Iguanodon, a three-ton herbivorous dinosaur half the length of a tennis court, appeared as a bloodthirsty, cannibalistic dragon in one image, an impish, overgrown lizard in another, an affable scaly elephant in the next," writes Zoë Lescaze in a brilliant history called *Paleoart*.

Artists and naturalists before the nineteenth century had even less to go on and came up with even odder images. For more than four hundred years, for instance, a six-ton sculpture of a fierce dragon—complete with wings and powerful tail and gaping jaws—has stood in the town of Klagenfurt, Austria.

The sculpture was based partly on the discovery of an imposing skull in 1335, in a local quarry. It is, by many accounts, the first-ever attempt at reconstructing an extinct animal.

For centuries, people in Klagenfurt had told and retold stories about a dragon that lived in the marshes at the edge of town. As townspeople recounted the tale, the creature emerged from mist and fog to devour travelers crossing the nearby river (*Klagenfurt* means, roughly, "the ford of lament"). Finally, brave knights lured the monster into the open by chaining up a bull as bait. When the dragon leapt on the bull, the knights burst out of hiding to slay it, thus saving the town.

In the 1300s, when a huge, fearsome skull turned up in the quarry, the legend seemed confirmed beyond any possibility of doubt. For hundreds of years after its discovery, the skull sat on display at the town hall. Eventually, in 1590, a local artist carved the dragon sculpture; the fossil skull served as his model for the dragon's head.

The skull can still be seen in the State Museum in Klagenfurt. It turns out to have come not from a dragon but from an Ice Age woolly rhinoceros, a formidable animal even though it had no wings and no taste for human flesh.

Cultures around the world have always treated bones, even run-of-the-mill ones, with care. Bones inspire awe and reverence, presumably because they stir thoughts about death and what remains after a life ends. *Gigantic* bones were both treasures and natural wonders, and they featured as hallowed objects in temple shrines and important buildings.

In Europe in the Middle Ages, for instance, churches and city halls and castles often displayed huge, mysterious bones. Decoration and veneration were close cousins. "A principal courtyard of Whitehall Palace, designated Whalebone Court because of the display of whalebones there, was one of the sights of London," one historian writes, "and the large bones of a rhinoceros, a whale, and a mammoth were suspended above the portal of Wawel Cathedral in the Royal Castle in Cracow."

Such displays continued into relatively modern times. "As late as 1789," the historian Claudine Cohen observes, "in praying for rain, the canons of Saint Vincent's would parade through the streets and countryside carrying what was believed to be a saint's arm but in fact was the femur of an elephant."

Every culture has made sense of giant bones in its own way. Until the 1800s no one dreamed of dinosaurs, because there was no need. Not when myth and legend played so large a role in everyone's thinking. Surely these huge, uncanny relics came from human giants or unicorns or hybrid creatures with sharp teeth and fearsome claws or—a global favorite—dragons.

In India, oversized bones were the remains of flesh-eating demons called rakshasas who had the power to change shape and appear as animals or as monsters.

In Europe, stories of human giants were particular favorites. In a book written around the year 100 AD, a Roman historian told of an immense earthquake during the reign of the emperor Tiberius. In the aftermath, a gash in the earth revealed a hoard of giant bones.

These were, the locals decided, the remains of bygone heroes. They took care to disturb the bones as little as possible, but they sent one immense tooth to the emperor by way of tribute.

Tiberius, who was eager to know the ancient hero's stature but unwilling to desecrate his grave, devised a plan. He assigned a mathematician to calculate the size of the skull that would have housed so impressive a tooth. Then he had a sculpture built to those dimensions. The tooth itself was returned home and reburied with proper ceremony. (Modern scientists guess that the tooth, which measured about a foot in length, came from a woolly mammoth.)

Tiberius was not a notable figure (except temperamentally, per-haps—the Roman historian Pliny the Elder called him "the gloomiest of men"), but some of the greatest names in the history of Western thought believed in giants just as wholeheartedly as he did.

Saint Augustine devoted long passages of his most important work, *The City of God*, to accounts of giants. "I myself, along with some others, saw on the shore at Utica a man's molar tooth of such a size, that if it were cut down into teeth such as we have, a hundred, I fancy, could have been made out of it. I believe it belonged to some giant."

How could such a thing be? Because when the world was new, Augus-tine explained, humans were much bigger and lived much longer than they did in the sinful, fallen world of his own day.

The Bible said so, and here was evidence to quiet any doubters. "The bones which are from time to time discovered," Augustine wrote, "prove the size of the bodies of the ancients, and will do so to future ages."

The giant tooth, which was brought to a church and displayed to the faithful, probably belonged to some sort of extinct elephant.

The New World had its own giant bones and its own stories about how they had come to be. In 1519, Hernan Cortez and his conquistadors landed in Mexico. Almost at once they heard about gigantic bones from the Tlaxcala, rivals of the Aztecs, who explained that their ancestors had managed to kill the evil, oversized giants who had once lived near them.

"So that we could see how huge and tall these people had been," one of Cortez's soldiers reported, "they brought us a leg bone of one of them, which was very thick and the height of a man of ordinary stature, and that was the bone from the hip to the knee. I measured myself against it, and it was as tall as I am, though I am of fair size."

The Spanish, who had come to the New World expressly to plunder, snatched the bone as if it were a lump of gold and put it aboard the next treasure ship bound for the king. The bone, which has since been lost, likely came from a mastodon.

Colossal bones weren't the only prehistoric relics that inspired thoughts of monsters and giants and dragons. In China over the course of recent centuries, generations of villagers in Sichuan Province passed down a legend about eighteen huge tracks that led up Luoguan Mountain.

The footprints, scientists now know, were made 150 million years ago by seventy-ton dinosaurs called sauropods. (Brontosaurs—immense, long-necked, small-headed creatures—are the best-known sauropods. Picture the hero of *Danny and the Dinosaur*.)

According to local lore, the tracks had been made by Divine Lucky Rhinoceros, which had set off in search of a particular red mushroom with magical healing properties. (Rhinoceroses roamed across China until around 1000 AD, when hunters killed the last one.) Legend decreed that good fortune came to those who followed the prints up the mountain.

Europe had telltale footprints of its own. In medieval Germany, legends told of how Siegfried, a mighty warrior, had killed a dragon that he had tracked to its lair by following a trail of immense prints pressed deep into the ground. The dragon was so huge that the ground trembled when it walked.

Perhaps those tales rose entirely out of a storyteller's imagination. But paleontologists have now found *three* different sites in Germany where enormous dinosaur footprints—the biggest are three feet across and two feet deep—march across stony ground.

Looking into Medusa's Eyes

In modern times, one scholar has done more than any other to probe the art and legends of past cultures to see if they might conceal dinosaurs or other prehistoric creatures. She is Adrienne Mayor, a Stanford University historian. It was Mayor who unearthed the stories of Divine Lucky Rhinoceros, in China, and Siegfried, in Germany, and dozens more besides.

She began with close looks into Greek and Roman legends. Paleontologists know bones. Classicists know Greek and Roman literature. Mayor knows both. That makes her an unlikely hybrid, which is perhaps fitting for an expert on the culture that produced Pegasus and centaurs and the Minotaur.

She had at least one notable predecessor. In 1914 an Austrian paleontologist named Othenio Abel looked at the skulls of extinct, smaller-than-present-day elephants—which turned up fairly often in digs in Greece and other sites around the world—and pointed out something odd.

Abel began with a straightforward description. Elephant skulls don't look like ours. Instead of two big, round eye sockets prominently positioned at the front of the skull, the eye sockets in an elephant's skull are at the sides and almost hidden. If you didn't know where to look, you could miss them altogether. Look at the *front* of an elephant skull, and you see a single giant opening, where the trunk attaches.

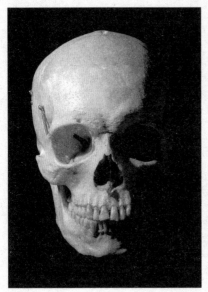 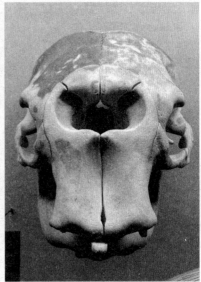

Then came the payoff. Ancient Greeks found those strange skulls, Abel suggested, saw one gaping hole where they expected two distinct eye sockets, and thought . . . *Cyclops*! (The other bones in a mammoth's body—ribs and femurs and so on—look more or less like oversized versions of their human counterparts.)

This was speculation, not proof, for no one has ever found an ancient document that declares, "Today I found a Cyclops skull." And even if Abel's claim was correct, it would not settle the chicken-and-egg question of which came first, the Cyclops myth or the elephant skull. Did the skull give rise to the myth, or did the myth make for a misreading of the skull?

Adrienne Mayor took up Abel's quest and vastly expanded it. One of the great scientist-detectives of recent years, she has made a specialty of looking at myths and legends from pre-scientific cultures and seeing if those ancient tales have any connection with modern-day scientific discoveries.

Mayor rose to fame in the 1990s with a claim about griffins, not Cyclopses. Griffins were hybrid creatures—they had the body of a lion and the beak and claws of an eagle—and they guarded stashes of gold. Greek and Roman writers described them often.

But in contrast with winged horses, say, griffins never turned up in Greek mythology. "Unlike the other monsters who dwelled in the mystical past," Mayor writes, "the griffin was not the offspring of gods and not associated with the adventures of Greek gods or heroes. Instead, griffins were generic animals believed to exist in the present day; they were encountered by ordinary people who prospected for gold in distant Asia."

Scholars had always regarded such stories as legends from the superstitious past. Mayor was less dismissive. She discovered tantalizing snippets in ancient manuscripts, like one from a Greek physician named Ctesias. In 400 BC he wrote about why gold from Asia was particularly hard to obtain.

His tone was surprisingly matter-of-fact, as if he were describing a flesh-and-blood animal rather than a mythical half-man/half-bull creature like the Minotaur. Though there was a great deal of gold in Asia's mountains, Ctesias wrote, it came from "an area inhabited by griffins, a race of four-footed birds almost as large as wolves and with legs and claws like lions."

Mayor found that similar accounts recurred over the course of a thousand years of Greek history. Many came with illustrations. Other scholars shrugged, if they took any notice at all.

Then a series of fossil-hunting expeditions in the Gobi Desert, first in the 1920s and then more extensively from the 1960s onward, turned up hundreds of skeletons from a dinosaur called protoceratops. The finds came from the same regions where ancient miners and nomads had found gold millennia ago.

The skeletons would have been impossible for those early witnesses to have missed. Unlike most fossil bones, which are dark and concealed by dirt and rock, these bones were creamy white and scattered atop the ground. They were "strewn over the surface almost as thickly as stones," the adventurer and naturalist Roy Chapman Andrews reported in 1920.

Protoceratops was about six to eight feet long—roughly lion sized—with a sharp, fierce beak, as if from an oversized eagle. Mayor has photographed the fossil skeletons alongside ancient drawings. Protoceratops and griffins look like twins.

Mayor tells another surprising story of fossils and myths, this one based on a strange image that art historians long ago dubbed the "Monster of Troy." As Homer retold the story, the monster was a fearsome beast that could only be coaxed to stop killing the locals if the Trojan king sacrificed his daughter, Hesione.

A painting on a Greek vase from the sixth century BC, now in the Museum of Fine Arts in Boston, picks up the tale at the crucial moment. Hesione stands just in front of the monster, pelting it with rocks. Next to her stands Hercules, who has shown up to join the battle and has launched a flurry of arrows at the beast.

Some of the rocks have hit their target. One is embedded near the creature's eye, and a second is lodged deep in its mouth. One of Hercules's arrows has pierced the beast's jaw.

What's odd is the monster itself, which is depicted as a large, white skull emerging from a black blob. The other figures on the vase—Hercules and Hesione, a horse, geese—are well drawn. Not the monster. One art historian complains about its "shapeless, unworthy head," and another describes it as a "hideous white Thing."

Look again! But this time bear in mind that the legend is set on a stretch of the Turkish coast where fossils erode out of cliffs that rise up from the sea. "Instead of a poorly drawn sea monster peeking out of a sea cave," writes Mayor, "we suddenly perceive a monstrous animal *skull* poking out of a cliffside."

The suggestion is surely plausible. Mayor has paired the image of the monster on the vase with a strikingly similar look-alike. (This is the fossil skull not of a dinosaur but of a prehistoric giraffe, from Greece.)

The match is not perfect, but it is close. In Mayor's judgment, "the strange head appears to be the earliest artistic representation of a fossil discovery in antiquity."

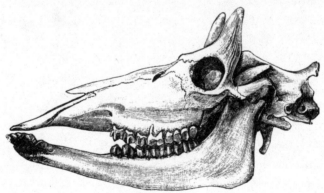

If Mayor has it right, creatures long thought to have lived only in folklore were once flesh and blood. More than that, they were flesh and blood that was transformed into stone.

It is a lovely coincidence—or perhaps a bit more than coincidence—that many Greek myths tell of living people and animals that were turned to solid stone. Medusa is the most famous—anyone who gazed into the

eyes of fearsome, snake-haired Medusa immediately turned to stone—
but in Greek storytelling the theme recurred often.

The ancient world never solved the riddle of fossils, but there was
something uncanny about the mystery, and they could never quite let
it go.

Did the Greeks have an inkling, centuries before scientists sorted it
out, that life and lifeless stone might somehow be linked?

Leibniz's Unicorn

As the Age of Science dawned in the West—in the 1600s, the era when Galileo first turned his telescope toward the heavens—thinkers who were grappling with the mystery of giant bones decided to change tack. Myth and conjecture were all very well, they declared, but now was the time for hard facts. Let us move beyond dragons and giants.

It was a noble plan but a hard one to put into practice. Figuring out what bygone animals had looked like was the biggest challenge, but even creatures far closer to home presented giant problems. Let's start with these "easy" cases and work our way along.

Even a generation or two after Galileo, the best educated guesses (and the best educated guessers) could still veer into fantasy. In a far-ranging volume on geology, for instance, one of the greatest names in the history of science reported one of the strangest of all discoveries.

In northern Germany, Gottfried Leibniz wrote, an extraordinary skeleton had been found, in 1663, in a limestone quarry. The skeleton, it seemed, came from a unicorn!

This was a startling claim, but Leibniz boasted impeccable credentials. He was the peer and rival of Isaac Newton and one of the most acclaimed thinkers who ever lived.

Brilliant and impossibly learned, he inspired awe and envy in every-one who crossed his path. "In the same way the ancients could manage simultaneously up to eight harnessed horses," one contemporary mar-veled, "Leibniz could manage simultaneously all the sciences."

No one had a more powerful mind. No one excelled in as many different fields. In the late 1600s Leibniz took a deep dive into geology. (This wasn't the plan. Leibniz had been hired to write a history of the Hanoverian dynasty in Germany. But his approach was always to bite off more than a dozen men could chew. He decided to begin his royal history at the beginning, with the formation of Earth.)

A unicorn sighting seemed completely unlikely, Leibniz happily conceded, but an investigator's task was to go and see, not sit at home and issue decrees on just what wonders the world was allowed to con-tain. Leibniz pointed out that a prominent local official, one Otto von Guericke, had vouched for the unicorn discovery. Von Guericke was not only the mayor of the city of Magdeburg, Leibniz noted, but also a well-regarded scientist who had "ennobled our era by his discoveries."

Von Guericke had a dramatist's flair, and he'd grown famous before he ever set eyes on a unicorn. His big moment came on a spring day in 1654, in front of an audience that included a host of prominent officials and Emperor Ferdinand III. Von Guericke had prepared an elaborate demonstration. First he took two identical, hollow hemispheres—imag-ine two Weber grills—and sealed the two halves together (with grease) to form a single large, hollow sphere. Then he pumped out the air from inside the sphere and created a vacuum.

While the emperor craned his neck for a better view, von Guericke harnessed a team of thirty horses, fifteen on each side, to try to pull the hemispheres apart. They snorted and strained and tore up the ground, but they failed.

Here was experimental proof of a startling fact—everyone on Earth lives inside a sea of air, as fish live inside a sea of water. That air exerts a great deal of pressure. We don't feel it because the pushes come from every direction and cancel one another out.

But if you devised a way to remove a push from one direction—as von Guericke did by removing the air from inside his sphere—then suddenly you'd see what happens when you have a push with no pushback. Even a team of horses would be outmatched.

Nearly four centuries later, the demonstration is still famous. Von Guericke was a figure to be reckoned with. And Leibniz, even more so.

So, when the two men combined forces—when Leibniz passed along the news that von Guericke had seen "a skeleton of a unicorn animal"— everyone paid attention. Especially when Leibniz provided a drawing of the unicorn.

It's hard to imagine a more bizarre creature. Presumably it tottered along on two legs. In comparison a camel looks like an embodiment of architectural grace. But, perhaps hypnotized by the spearlike horn, no one protested.

The creature's appearance is so strange because its skeleton had not been dug up intact. The bones had been "extracted by pieces, because of the ignorance of the diggers," Leibniz explained, and then von Guericke had fit the pieces together.

Modern-day scientists believe the "unicorn" was in fact a hodgepodge made from at least two different extinct animals. The skull belonged to a woolly rhinoceros and the teeth and legs to a mammoth. The spine in the drawing is upside down and backward (the "ribs" were actually vertebrae).

The tusk is harder to identify. It doesn't spiral like a narwhal's tusk, and it's too long to have come from a walrus. One modern paleontologist suggests that it might have come from an extinct elephant that had straight tusks.

The unicorn story is noteworthy not because Leibniz and von Guericke went wrong, but because it highlights how immensely difficult it was to reconstruct unknown animals from a collection of bones. Unicorn-style blunders were nearly inevitable.

Faced with incomplete information, there was no choice but to fill in the blanks as well as you could. And, after all, no one had an actual unicorn to serve as a model.

Put unicorns and dinosaurs to one side for a moment, and look at the difficulties that confronted early naturalists even when they tried to depict genuine animals of their own day. If the animals were unfamiliar or exotic, all bets were off.

Look at this drawing of a whale. It comes from an encyclopedia that historians praise as a landmark in the history of science. The whale had beached itself at the mouth of the Thames, in 1532, and drawn a curious crowd.

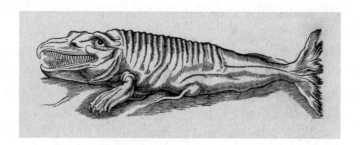

Here is another whale from the same encyclopedia, the Swiss naturalist Conrad Gessner's *History of Animals*. The image shows a whale bearing down on a hapless ship.

These whale images have an awkward charm, but clumsiness on the artist's part isn't the reason the drawings are off. Look at Albrecht Dürer's famous—and famously fanciful—woodcut of a rhinoceros from 1515. Dürer was as skilled a draftsman as ever lived, but he had never seen a rhinoceros.

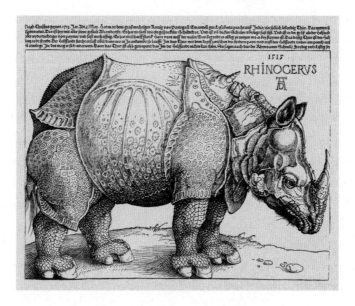

He relied, instead, on rumors and on a sketch by an artist who actually *had* been granted a peek at one of the remarkable beasts. Such glimpses were astonishingly rare. Fifteen centuries before Dürer, Roman emperors had managed a few times to bring rhinoceroses to the Colosseum, where roaring crowds watched them killed with spears or bows and arrows.

In the millennium and a half that followed, virtually no Europeans laid eyes on a rhinoceros. The huge beasts faded from Western memory altogether or wended their way, mythically transformed, into fables and legends. Then, come the Renaissance, they lumbered back onto the European scene.

In Dürer's day, exotic animals were deemed the perfect gift for the man who had everything. In 1515 the sultan of Cambay (in the west of India) sent a rhinoceros to the king of Portugal. In short order the king regifted the rhino to Pope Leo X.

Leo, a member of the immensely wealthy Medici family, was a hard man to shop for. He was a complex character, learned, hedonistic, and unscrupulous in roughly equal measure. (It was Leo who supposedly said, "Let us enjoy the papacy since God has given it to us.")

For centuries the church had raised money by selling indulgences, which were payments to have one's sins forgiven. Leo came up with a new and hugely lucrative twist—payments for the forgiveness of *future* sins—and in so doing inspired Martin Luther's historic protest against Rome's corruption.

The rhinoceros gift was a redo. The king of Portugal had sent Leo an elephant the year before, but it had died shortly after its arrival. The rhinoceros fared even worse. The ship carrying it to Rome ran into a storm just off the Italian coast. The rhinoceros, chained to the deck, went down with the ship.

Dürer based his woodcut on a sketch an artist had made during the rhino's brief stint in Lisbon. The image is an artistic triumph, but it is a mélange of imagination and guesswork.

Dürer dreamed up a suit of elaborate, layered armor for the beast, complete with a small breastplate, and he added a second, spiraling mini-horn at the base of the neck. Nonetheless, the historian Sue Prideaux remarks, "the famous woodcut stood as rhinoceros-truth for two hundred years."

Ponder these two-legged unicorns and grinning whales and their unlikely brethren for a moment, and you see why early scientists and artists had so much trouble trying to imagine what *dinosaurs* looked like.

Every prehistoric animal was a mystery, but dinosaurs were by far the closest to a blank slate. (Everyone knew that unicorns looked like horses or deer.) Naturalists in those pioneering generations were on their own, stumbling along and looking for precedents, even far-fetched ones, wherever they could find them.

"For people in the early nineteenth century," the historian Ralph O'Connor observes, it was all but impossible "to imagine these new monsters without drawing on old iconographies."

They began with the Bible. No images were as well-known as those in the Bible. There was no mention of prehistoric creatures in scripture, of course,* but the Bible *did* tell stories from the earliest days of creation, and religious texts were chockablock with pictures and descriptions of animals.

These tended toward the cheerful and peaceful—lions lying down with lambs, deer cavorting in the Garden of Eden. (According to Christian doctrine, even the lions in Eden were vegetarians. The reason Eve was not afraid of the serpent, Augustine explained, was that all the creatures in the garden were harmless.)

So the earliest attempts to depict dinosaurs had a strangely biblical flavor. Smiling dinosaurs gazed out from tranquil riverbanks. But

---

* Nearly all the animals that turn up in the Bible are familiar, with the notable exception of Leviathan. Today people tend to assume that Leviathan was a whale, but this was no ordinary whale. Leviathan is a "crooked serpent" and "the dragon that is in the sea," we read in Isaiah, and in Job we hear of his "scales" and "terrible" teeth and learn that "a flame goeth out of his mouth" and "out of his nostrils goeth smoke." We will encounter Leviathan again in chapter 29.

peaceful scenes quickly gave way to violent ones, which seemed to speak to something deep in the human psyche.

And despite the earnest efforts of early scientists to leave the past behind, old myths retained a stranglehold on their imagination. O'Connor, the historian, cites a painting of a pterodactyl from 1829. "The pterodactyl is not represented *like* a dragon," he notes, "but *as* a dragon."

"The Grinders of an Elephant"

Sorting out what dinosaurs looked like was so hard because there was too much leeway, too much room for imagination to run free. But by the nineteenth century, scientists were after bigger game than mere description. They wanted to understand the prehistoric world in general, and that meant tackling a slew of *who? when? where?* mysteries.

They rushed boldly ahead, certain that they were on the brink of success.

They had no idea.

The problem was that bones were the crucial clues, and when it came to bones, nothing was simple. Time and happenstance had broken them apart and flung them together in dozens of head-scratching ways, as if a horde of passersby had tramped through what should have been a roped-off crime scene.

Scientists in the early 1800s had not yet seen the big picture. Stories that should have been utterly distinct were jumbled together. The result was a kind of chronological traffic jam, as in the sort of children's books where a cowboy on horseback throws a lasso around the neck of a runaway brontosaurus.

By the early decades of the nineteenth century, so many immense bones had turned up that it was impossible to doubt that huge, mysterious animals had once roamed the earth. But what those creatures looked like and when they had lived, no one yet knew. No one even knew if they were *still* alive.

No one had seen them lately, it was true, but perhaps they had not looked in the right places. Every continent was dotted with blank spaces, and the oceans were hardly anything *but* blanks. Except for what hints you could gather by peering overboard or lowering a net a short way into the bottomless depths, no one had any idea what the seas concealed.

When you dug up an ammonite, say, it seemed possible that you'd happened on a creature that had wandered far from home, not that you'd found a relic of an ancient world. Perhaps ammonites were lost in space, not lost in time.

So the first item on the scientists' to-do list was sorting out who had lived when, and where, and whether they were long gone or just in hiding. Whose bones were whose?

Stashes of bones seemed to turn up everywhere you looked. In present-day London, for instance, every tourist and every local knows Trafalgar Square. It is home to the National Gallery and to Nelson's Column, which rises high into the sky with the naval hero at its summit and four carved lions at its base. At every political demonstration, protesters by the thousands gather in the square to wave signs and banners. It is, in short, a quintessentially modern, urban site.

And yet, when workmen began tearing up the area in the 1830s, just before Queen Victoria came to the throne, they found heaps of bones that scientists tentatively assigned to rhinoceroses, elephants, hippos, hyenas, and saber-toothed tigers.

Those creatures lived in the most recent ice age, we now know, which means they came along tens of millions of years after the dinosaurs vanished. Then a great many of them vanished, too, ten thousand years ago.[*]

---

[*] Most of the great beasts died out at about the time humans turned up. That doesn't sound like a coincidence. But even though hunting was certainly a giant part of the story, it was apparently not all of it, and scientists have yet to sort out how climate change and other factors came into play.

That picture was starting to come into focus in the 1800s, but important features of the landscape were still blurry. Scientists did know by this point that rocks formed thick layers underground and that lower layers were older than ones nearer the surface. That much was well established.

Geologists dignify this notion with a grand name—the "principle of superposition"—but the idea is familiar to anyone who has seen a teenager's bedroom, where the T-shirts and discarded underwear on top of the heap on the floor are plainly more recent than the debris in the lower layers. (In practice, matters are not so simple, because Earth's crust is always in restless motion. We need to imagine that a golden retriever puppy has just dived into the dirty clothes in search of a tennis ball.)

Saurian fossils turned up in layers deeper than mammalian fossils. So dinosaurs were older than woolly rhinos and saber-toothed tigers. Scientists knew that much.

But how much older? No one knew. They *thought* they knew that dinosaurs and woolly rhinos had died out long ago, but a few holdouts doubted even that. Was there some remote jungle where dinosaurs still reigned or woolly mammoths shook the earth?

When it came to muddling up past and present, mammoths and mastodons were the chief culprits. It was their elephant-like skulls and tusks that made them so hard to classify. Were those bones and teeth relics from present-day animals that had made their way to new homes, or did they come from ancient beasts that had died long ago?

Some of the most puzzling stories came from Russia, which in this era almost embodied the unknown and unexplored. In 1721, a delegation of Russians traveled to China, where they passed along astounding news. One region of their country, they told their stunned listeners, was so cold that the ice there never melted!

Deep under the ice and snow, the Russians went on, they had found "a rodent similar to an elephant, which makes its way underground and which expires the very moment it is exposed to light or air."

To modern ears, a burrowing mole as big as an elephant, frozen underground in Siberia, sounds distinctly like a long-buried woolly mammoth. The Russians had reached a different conclusion.

The giant moles were present-day animals, they believed, not relics of a bygone age. The reason they were never seen alive was that they were not hardy enough to withstand exposure to daylight and the open air.

In America, gigantic bones and teeth had turned up at virtually the same time as in Russia, but they inspired a vastly different story. The trouble started in 1705 when a farmer near Albany, New York, found what a local newspaper called "a great prodigious Tooth."

The strange object quickly moved its way up the social ladder. The Albany farmer bartered his find away for half a cup of rum. The new owner passed it on to Lord Cornbury, the governor of New York. Cornbury shipped it to the Royal Society in London, with a note describing it as "the tooth of a Giant."

More digging near where the Albany tooth had turned up yielded a stash of enormous bones. Based on their size, one early eyewitness wrote, "the Monster was judged above 60 or 70 feet high."

Both the bones and tooth were baffling. "Some said 'twas the tooth of a human creature; others of some beast or fish," Cornbury told the Royal Society, "but nobody could tell what beast or fish had such a tooth."

Most people shared Cornbury's view that this was the tooth of a human giant. The idea was that enormous teeth and bones were relics of giants who had drowned in the biblical flood. The Bible was gospel, as it were, and here was proof enough to gladden the heart of any believer.

Leading thinkers hurried into print. Cotton Mather, the renowned Puritan minister, wrote excitedly about the huge tooth. He had seen and handled it himself, he boasted, and he was starstruck. Here was an "Illustrious Confirmation" of the Bible's talk of giants.

Better yet, Mather exulted, here was a New World marvel to wave in the face of old, smug Europe. Even when it came to disputes over who had better giants, the upstart American colonies could hold their own against ancient lands. *"Men,* who were able to *turn the world upside down,* came hither also."

Mather was one of the best-known Americans of his day. His opinions carried weight, although in all his judgments he managed the neat trick of looking to the future with one eye and to the past with the other. Mather was a member of the Royal Society and also an earnest believer in black magic. He argued that inoculations against smallpox would save countless lives,* *and* he was one of the great villains in the Salem witch trials.

That jumble of ideas, with reverence for science joined with an embrace of superstition, was typical of the era. And in his faith in giants, Mather was an emblem of his age, not an outlier.

In the meantime, bones and teeth strikingly like those from the "Monster" began popping up all across America. The crucial clue came from South Carolina, in about 1725. There a group of enslaved people—their names have vanished from history—found several huge teeth in a swamp.

Plantation overseers delivered the usual verdict—these were relics of human giants. The Africans who had made the find had a different theory. They recognized these teeth. They looked like molars they had seen in their homeland, and they weren't human. These were the teeth of an elephant.

That was correct, or, at least, nearly correct. The first to hail the breakthrough was an English naturalist and illustrator named Mark Catesby who saw the newly discovered teeth while on an extended trip to the American South.

Catesby was open-minded and inquisitive (he was one of the first to solve the age-old mystery of where birds go in winter). "At a place in

---

* Mather heard about inoculation from an enslaved African man named Onesimus who had been presented to Mather as a gift by his church congregation. Onesimus had learned about inoculation in West Africa. He explained the procedure in detail. "In Africa," Mather wrote, "where the poor creatures die of the Small-pox like Rotten Sheep, a merciful God has taught them an Infallible Preservative. 'Tis a common practice, and is attended with a Constant Success."

Carolina called *Stono*," he wrote years afterward, "was dug out of the Earth three or four Teeth of a large Animal, which by concurring opinion of all the *Negroes*, native *Africans*, that saw them, were the Grinders of an Elephant."*

Catesby knew the identification was correct, because he'd seen elephant teeth in London. "In my opinion they could be no other," he wrote, "I having seen some of the like that were brought from *Africa*."

That was in 1743. Almost exactly two centuries later, in 1942, an eminent American paleontologist grudgingly acknowledged Catesby's account. "It appears that Negro slaves made the first technical identification of an American fossil vertebrate," wrote George Gaylord Simpson, "a lowly beginning for a pursuit that was to be graced by some of the most eminent men in American and scientific history."

At this point the story took an odd turn. Before it ended, it would embroil the upstart United States against proud, disdainful Europe in a battle for national prestige. Benjamin Franklin would wade into the fray, and Thomas Jefferson would take a leading role.

---

* Stono is an important name in American history because one of the biggest uprisings of enslaved people took place nearby, about a decade after the fossil find. On September 9, 1739, a group of enslaved people gathered near the Stono River in South Carolina and began marching south, under banners reading *Liberty*. (The Spanish had recently announced that they would give freedom and land to anyone who reached Spanish Florida.) By dusk the group had grown to around sixty. White pursuers killed about half of the rebels, and about half escaped. Twenty or twenty-five whites were killed in the fighting. Nearly all the rebels who survived the day's combat were soon captured and executed.

"The Terror of the Forest"

At the center of the fray stood the "elephants" whose giant molars had turned up in a South Carolina swamp.

The teeth from those towering beasts were found well before Pliny Moody's dinosaur footprints or Mary Anning's fossils. But the New World "elephants"—they would eventually be identified as mastodons—played an important role in the dinosaur story even so, because they spotlighted the question of extinction.

When it came to prehistoric life in general, and to dinosaurs in particular, the single most important, most perplexing question of all was, *what happened to them?* And just as the mastodon discoveries foreshadowed the dinosaur discoveries, so the mystery of the mastodons' whereabouts foreshadowed the mystery of the dinosaurs' whereabouts.

By the late 1700s, the era of Ben Franklin and Thomas Jefferson, American farmers tilling the soil and workmen digging ditches had managed to unearth entire skeletons from these almost-elephants. American patriots immediately seized on the finds as proof of the new nation's power and dynamism. Here were fitting symbols of a rising republic. In comparison with America's mighty pachyderms, what was the British lion but a forlorn tabby?

Pundits and public figures competed to paint ever more fearsome images of the creature, which was dubbed the American incognitum (Latin for "unknown"). Later it would be identified as a mastodon, although most people continued to use the word *mammoth* rather than the scientific name.[*]

By any name, this was a formidable creature. "Forests were laid waste at a meal," one writer declared in the 1790s. And not just forests, for unlike present-day elephants, these immense creatures were, supposedly, meat eaters. They "especially loved to feed upon human flesh," one expert declared, and another claimed that "whole villages, inhabited by men, were destroyed in a moment."

A description from 1797 went further. "With the agility and ferocity of the tiger, with a body of unequaled magnitude and strength, it is possible the Mammoth may have been at once the terror of the forest and of man!"

These reports of a huge, fierce, and, above all, *American* creature were welcome news because European scientists had proclaimed for decades that such a thing could never be. America was an inconsequential country, and its animals were small and stunted, especially in comparison with the behemoths that had once ruled the Old World.

The spokesman for the European camp was a vain, brilliant, flamboyant French scientist named Georges-Louis Leclerc, a sneer in human form. Count Buffon, as he was known (the name honored his hometown), had inherited a fortune when he was young and had been granted a title by the king.

Buffon was as far as could be from the stereotype of the ascetic, unworldly scholar. "He loved money and became rich," wrote Stephen Jay Gould. "He loved power, and he frequented those in power. . . . He loved women, and not just for their beautiful souls."

No one disputed Buffon's intellectual powers, least of all Buffon himself. He had special regard, he wrote, for a tiny group of thinkers

---

[*] Mastodons and mammoths were different creatures—mastodons appeared far earlier in history and were smaller—but in popular usage *mammoth* applies to both.

who stood atop the world. "There are a bare five. . . . Newton, Bacon, Leibniz, Montesquieu, and me."

The boast had a basis in fact. Buffon truly was one of the towering figures of eighteenth-century science. He was not only the best-known naturalist of his day, but—in a display of versatility that was a feature of his era but would be almost impossible today—he was a major contributor to mathematics and astronomy as well.

He rose to fame on the basis of an immense thirty-six-volume work called *Natural History*. In lively and opinionated prose, Buffon ranged across the world and around the heavens, taking on everything from astronomy to botany to geology to zoology.

No topic was too big; Buffon explained the origin of the solar system. No topic was too small; Buffon reviewed the natural world animal by animal and plant by plant, handing down verdicts like a cranky judge: "Sloths are the lowest form of existence in the order of animals with flesh and blood; one more defect would have made their existence impossible."

The volumes of his *History*, issued year after year across the decades, made Buffon famous throughout Europe, with a reputation that extended far beyond literary salons and academic institutions. When he died, in 1788, twenty thousand spectators lined the streets to witness his funeral procession. The French Revolution would erupt only a year later, but the bewigged nobleman could not have had a grander send-off. Fourteen elaborately outfitted horses led the procession. Then came nineteen servants, sixty priests, and thirty-six choirboys.

Buffon never visited North America, but he delivered harsh judgments on the New World nonetheless. It was America's misfortune to suffer from a cold, damp climate, he explained, and those dismal conditions had "shriveled and diminished" American animals, leaving them smaller and weaker than their European counterparts.

Buffon cited the American lion, or puma, which fell a long way short of the king of beasts. It lacked a mane, and that was only one shortcoming. As if anticipating *The Wizard of Oz*, Buffon rattled off the puma's other deficiencies: "It is also much smaller, weaker, and more cowardly than the real lion."

Similarly, the Old World had elephants, but the New World could manage nothing more imposing than the tapir, a large, herb-eating mammal with a long snout. Buffon jeered, "This elephant of the New World is the size of a six-month-old calf, or a very small mule."

In America, one of Buffon's allies observed, dogs could not summon the strength to bark. Birds were too listless to sing.

New World plants and trees were puny and frail, too. So were Native Americans. "In the savage, the organs of generation are small and feeble," Buffon wrote. "He has no hair, no beard, no ardour for the female. . . . Their heart is frozen, their society cold, and their empire cruel."

Thomas Jefferson leapt up to make the case for the defense. Motivated by two of his great passions—love for America and love for science—he tore into Buffon's arguments with fury. The incognitum served as his smoking gun.

The incognitum was almost unimaginably huge, Jefferson wrote, "five or six times" bigger than an elephant. "It is certain such a one has existed in America, and that it has been the largest of all terrestrial beings," Jefferson thundered. *How dare Buffon!* The mere thought of so mighty a beast should have sufficed "to have stifled, in its birth, the opinion of a writer the most learned of all others in the science of natural history."

Jefferson had been obsessed with the incognitum for decades. In 1781 he'd written a note to his old friend George Clark asking him to make a trip to the Ohio River to collect incognitum bones. (George Clark was the brother of William Clark, of Lewis and Clark fame. The incognitum story served as a sort of real-life *Forrest Gump*, and every big name of the era popped up sooner or later. The messenger who carried Jefferson's note to Clark was Daniel Boone.)

Fascination with the incognitum was not an idiosyncrasy that set Jefferson apart from the other founding fathers. Ben Franklin had pored over a set of "elephants' tusks and grinders" that had been collected in the Ohio River Valley. Franklin was living in London at the time, on business representing Pennsylvania. He eagerly showed his collection—four tusks, four molars, and a bit of backbone—to English and French scientists, and together they puzzled over the strange fossils.

George Washington was intrigued by the incognitum, too, even when he had pressing business demanding his attention. In the winter of 1780, when the Continental Army was in winter quarters north of New York City, Washington heard rumors that some incognitum bones and teeth had turned up on a nearby farm. Along with a few fellow officers, Washington took a sleigh ride to see for himself. He examined the fossils with care and told the farmer about his own elephant "grinder," which he had been given as a gift.

But Jefferson's interest in prehistory ran deepest. He personally financed expeditions in search of fossils, and he kept a collection of incognitum bones on display in the entrance hall at Monticello, his plantation. During his presidency he piled up mammoth bones in the East Room of the White House.

Then, in 1797, Jefferson found an even better argument against Buffon than the incognitum. A friend named John Stuart sent him some huge fossil bones that had been found in a cave in what is today West Virginia. They were "the Bones of a Tremendious animal of the Clawed kind," Stuart wrote, and he believed they were "of the Lion kind."

In the early 1800s, the natural world was a safe and cozy place, or so people believed. "It is a happy world," declared the philosopher William Paley (at left, in a painting by George Romney), one of the leading thinkers of the age.

The idea of dinosaurs had never crossed anyone's mind. The only animals anyone could imagine were those from the Garden of Eden, as in this painting by Johann Wenzel Peter.

The Victorian age saw a frenzy of digging for railroads, roads, and canals. Workers unearthed huge, mysterious bones. Whose were they?

Giant bones had turned up through the centuries in many lands. They were displayed as relics worthy of reverence, but no one knew what they were. This fossilized bone—according to local legend it came from a dragon—has hung at the entrance to a cathedral in Kraków, Poland, for centuries. It is likely a whale bone.

Mary Anning was the greatest fossil finder of them all. She overcame three obstacles to a scientific career: she was poor, uneducated, and female. In this portrait she holds a geological hammer in one hand and points to an ammonite fossil (akin to a nautilus) with the other. Her dog, Tray, accompanied her on all her outings.

At right is Anning's first major find, an ichthyosaur, which roamed the seas 200 million years ago. At bottom is her second great find, a prehistoric sea creature called a plesiosaur. Its appearance was so strange that experts at first suspected it might be a hoax.

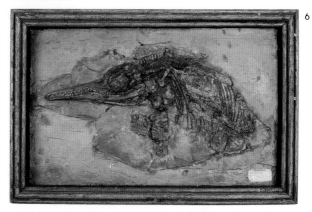

Museums eagerly displayed Mary Anning's fossil skeletons from the start, but her name went unrecorded. (Instead, museums credited the donor who had made the gift.) Today Anning's name features prominently near all her finds.

Anning lived all her life in the seaside town of Lyme Regis, on England's southern coast. Life was harsh and the Anning family struggled to get by. Lyme Regis today is prosperous and popular with tourists.

Lyme Regis is still rich with fossils that lie buried in its cliffs. Fossil hunting can be dangerous work. In 1833, Anning was nearly killed when a cliff crumbled and tons of rock crashed down near her. Her dog, Tray, was killed. "It was but a moment between me and the same fate," she recalled.

Gideon Mantell was a physician and an obsessive, brilliantly talented fossil hunter. Two of the first dinosaurs ever discovered were his finds.

Richard Owen coined the word *dinosaur*. He was the best-known scientist in England in the first half of the nineteenth century. Owen was a brilliant, complicated man with a gift for cultivating friends and making enemies. Mantell despised Owen and claimed that Owen had rewritten history to downplay Mantell's role.

William Buckland was one of England's most renowned geologists and perhaps the most eccentric Victorian of them all. (He lived in a household crammed with fossils and overrun with animals, and he served dinners that featured such delicacies as mice and crocodiles.) Here he holds a hyena's skull.

As late as the 1700s, even the greatest scientists believed that the Bible was the word of God, and its every word was true. But they struggled to make sense of the story of Noah's ark. How to reconcile the Bible story with the discovery of strange animals like kangaroos and possums, or with dinosaurs?

One of the crucial discoveries of the nineteenth century was that the world is ancient. Edmund Halley, of Halley's Comet fame, had been one of the first to see the truth back in 1715, but few people in his day paid attention. (Halley's insight rested on a fact that had been taken for granted for millennia: the ocean is salty. Halley thought to ask why.)

EDMVND. HALLEIVS LL.D.
GEOM. PROF. SAVIL. & R.S. SECRET.

London's Natural History Museum is a cathedral to science. It was Richard Owen's creation and his greatest legacy.

The outside of the Natural History Museum is adorned with animal sculptures. Owen disdained Darwin's theory of evolution, and he gave instructions to separate extinct creatures (on the east wing) from living animals (on the west).

17

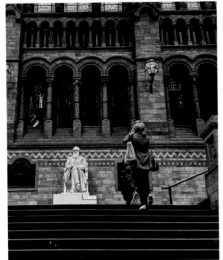

A statue of Richard Owen once stood in a place of honor in his museum. Today Darwin reigns unchallenged, and the Owen statue has been pushed aside in his favor.

Since no one has ever seen a dinosaur, no one can be certain what they looked like. Richard Owen believed they looked like oversized mammals and stood on massive, column-like legs. This model, built to his specifications, was one of a large group that dazzled nineteenth-century visitors to London's Crystal Palace Park.

18

19

Scientists today believe many dinosaurs sported feathers and stood upright. This is a model from the Vienna Museum of Natural History. It depicts deinonychus ("Terrible Claw"), a relative of velociraptor that stood approximately eleven feet tall.

"The Very Extraordinary SKELETON"

Jefferson delighted in the newest addition to his fossil collection. The claws—huge, sharp, and impossible to miss—were even more impressive than the bones. The claws of an African lion were a mere one and a half inches, Jefferson noted happily, while these claws of the "American lion" were *eight* inches long.

With the "victorious evidence" he had long sought now in hand, Jefferson prepared a paper on the colossal beast he had named megalonyx (from the Greek for "great claw"). In March 1797, he announced his find to the American Philosophical Society in Philadelphia. (Philadelphia was then the nation's capital, and Jefferson had traveled there not only to deliver a talk on fossils but also to take office as vice president.)

Megalonyx was "*more* than three times as large as the lion," Jefferson proclaimed. Who would dare look down on America now?

All Europe, it turned out, for megalonyx proved not to be a lion at all. It was, in truth, a slow-moving, plant-eating giant sloth.

This demotion—from the terror of the plains to a bumbling leaf chewer—was not welcome news. Worse still, Jefferson had learned the truth a few days *before* his talk but after he had sent off his manuscript. He could not bring himself to revamp his paper entirely. Instead he hedged his bets.

The trouble arose when Jefferson wandered into a Philadelphia bookstore and flipped through a copy of London's *Monthly Magazine*, which had just arrived. An article carried illustrations of a newly discovered, twelve-foot-long fossil skeleton from Paraguay.

The name of the new creature was megatherium (from the Greek for "great beast"), and it was identified as a colossal relative of the sloth. The giant sloth's bones and claws looked distressingly like those of Jefferson's mega-lion.

Jefferson raced to add a postscript to his paper. He had received new information "after the preceding communication was ready to be delivered," he wrote, and it appeared that a second creature had bones much like those of megalonyx. The newcomer's arrival on the scene *did not* mean he'd made a mistake, Jefferson emphasized, since more bones might turn up and show that megalonyx and the sloth were different creatures after all.

Even so, he raced to make changes to his paper. The fixes were subtle—Jefferson kept the passages that pointed out how much bigger megalonyx was than a lion, and he kept the tables comparing the length of the megalonyx bones with the corresponding bones in a lion, but he no longer claimed that megalonyx *was* a lion. (Wherever he had described megalonyx as a lion, he now substituted a generic reference to an animal "of the clawed kind.")

For Jefferson, the lion-versus-sloth controversy was major news. Science was a passion, not a diversion from his true calling; he sometimes regretted that he had chosen a career in politics instead. Natural history was the most compelling field of all, in his view, all the more so if he could find some feature of the world to count or measure or record.

Throughout his life, Jefferson was obsessed with data (the distance between cities, as measured by a newfangled odometer he had installed

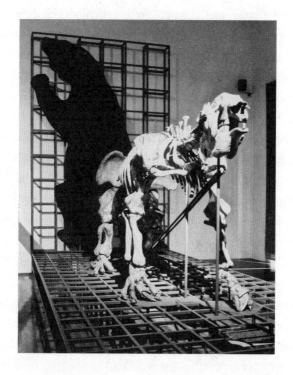

Skeleton of a megatherium, a giant ground sloth, as it was mounted at
the Royal Cabinet of Natural History in Madrid in 1789. It was the
first extinct mammal ever displayed in a museum. Modern scientists believe
the giant sloth stood upright to feed, as in the silhouette on the rear wall.

on a wheel of his carriage; the dates when thirty-seven different vegetables first appeared in Washington, DC, markets; the birds and plants he observed on his travels).* For more than fifty years he kept meticulous notebooks with daily records of the temperature. On July 4, 1776, a busy day for the founding fathers, he recorded the temperature at four different times.

So the possible demotion of megalonyx from lion to sloth was a serious blow. But Jefferson could always console himself with the incognitum, the best possible rebuttal to Buffon's "very degrading" theory.

The huge beast spoke for itself. Who needed data or Jeffersonian eloquence when the public could gaze on the breathtaking evidence with their own eyes?

In the fall of 1802—this was the same year that Pliny Moody found his mysterious footprints in Massachusetts—crowds in London squeezed their way into an exhibition hall, lured by newspaper advertisements touting "the very extraordinary SKELETON of the MAMMOTH."

Mammoth was an old word that had been newly imported from eastern Europe (it meant "earth burrower" and had been coined by medieval farmers who had found giant bones in their fields and guessed wrong about where they came from). *Mammoth* proved much catchier than *incognitum*. Soon the beast would be rebranded yet again, this time as a "mastodon," but the public liked *mammoth* best of all. Colonial America, especially, was knee-deep in oversized loaves of "Mammoth Bread" and wheels of "Mammoth Cheese."

The mammoth on display in London had been brought from America with enormous fanfare. It had been dug up (along with two others) the year before, on a farm in New York State's Hudson Valley.

Tickets cost the moon—two shillings, nearly two days' wages for a workman—but hardly anyone had ever seen so complete a skeleton. Museum exhibits at the time typically featured individual bones; this giant beast seemed ready to parade across the room.

---

* He was not above noting, as one counter to the theory of American puniness, yet another data point: he stood six foot two while Buffon was a shade under five foot five.

The skeleton had already dazzled audiences in the Philadelphia Museum, one of the first museums in the young republic. Now the museum's founder, an artist and showman named Charles Willson Peale, had sent his young son Rembrandt to England to win overseas admirers. (Rembrandt Peale had fifteen siblings, including a Rubens, a Raphaelle, and *two* Titians.)

The skeleton had two impossible-to-miss features. First was its size—at eleven feet high and seventeen feet long, it was about as big as an elephant. (In Philadelphia, as a sort of wink to museumgoers, Charles Peale had mounted a mouse's skeleton alongside it.)

Second, and just as striking, the skeleton had immense, downward-curving tusks. Those tusks had inspired endless debate—elephants' tusks turned *up*, but perhaps that was not fierce enough for a mastodon. Were the downward-curving fangs of a saber-toothed tiger a better model?

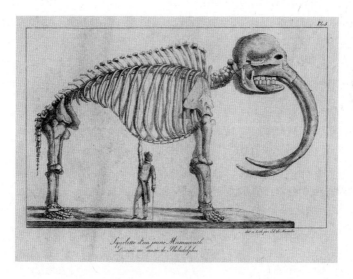

Squelette d'un jeune Mammouth.
Dessiné au musée de Philadelphie.

Rembrandt Peale considered both options and decided in favor of the tiger model. "In the inverted position of the tusks," he wrote, "he could have torn an animal to pieces held beneath his foot."

This was wishful thinking. Mastodons were vegetarians (the telltale clue was the shape of the molars, which were adapted for grazing on leaves and twigs). The skeleton's tusks had been put on upside down.

Jefferson firmly believed that extinction was impossible and that the unexplored regions of the world might conceal astonishing surprises. "In the present interior of our continent," he wrote in 1801, three years before he sent Lewis and Clark west, "there is surely space and range enough for elephants and lions, if in that climate they could subsist, and for mammoths and megalonyxes who may subsist there."

In 1804 he sent Lewis and Clark on their way with a long and detailed set of instructions. Among their many responsibilities, the explorers were to keep an eye out for "the animals of the country generally, & especially . . . any which may be deemed rare or extinct."

Somewhere in those vast plains, hulking mastodons and oversized lions might yet rule in splendor.

Noah's Ark

Jefferson's hope seemed far-fetched, but scientists had aspired to catalog and curate the natural world since the days of the Greeks. And
since the days of the Greeks, nature had seemed to delight in mocking
them for their presumption.

Every time it had seemed that *now* the world had offered up its best
and final surprise, another unlikely creature had popped into view and
marched (or sprinted or slithered or hopped) into the spotlight.

That long-running story seemed to have a simple moral: *You never
know!* Until the late 1700s, for instance, no European had ever seen
a kangaroo. (The first to make a sighting, the men of Captain James
Cook's *Endeavour*, promptly ate the surprising creature they had just
encountered.)

In ancient times, Europe had thrilled to its first peeks at exotic creatures
like elephants and lions. Then came the Age of Exploration, beginning
at about the time of Columbus. Descriptions of strange new animals
spilled out from explorers' journals like treats from a piñata—anteaters,
llamas, armadillos, hummingbirds, sloths, possums.

There seemed no end to the discoveries. In 1590, a century *after* Columbus, a Spanish missionary who had spent fifteen years in South and Central America flung up his hands in astonishment. He had seen "a thousand kinds of birds and fowls and forest animals that have never been known before either in name or shape," José de Acosta wrote, "nor is there any memory of them in the Latins and Greeks, nor in any nation of our world."

Historians have struggled to convey just how astonished Europeans were by this cascade of discoveries. (It's telltale that early writers referred to the Americas as the "New World" and not simply a "new continent.") Confronted with the completely unexpected, unfathomable wonders of America, writes Stephen Greenblatt, "European culture experienced something like the 'startle reflex' one can observe in infants: eyes widened, arms outstretched, breathing stilled, the whole body momentarily convulsed."

England was late to the game. While Spain and Portugal were sending home elephants, lions, parrots, and cheetahs from South America and Africa, one historian writes, "the English had only cod from Newfoundland." But eventually nearly every nation in the world joined the *Can you top this?* sweepstakes.

New animals were more exciting than anything else, but even fruit from far-off tropical locales seemed dazzling. Then greenhouses came along. They were hugely expensive to maintain, which meant that warm-weather treats like bananas and pineapples came to embody luxury and wealth.

In the mid-1700s, pineapples were so costly and so coveted that they were used mostly for display, not eating; pineapple rental shops sprang up to meet the demand. Prices stayed sky-high into the 1800s. As late as 1807 a man was sentenced to seven years' transportation to Australia for stealing seven pineapples.

Still, when it came to making an impression, nothing could compete with an unfamiliar animal from a distant land. Every new creature brought hordes of gaping admirers, from kings and queens in all their finery* to boisterous crowds pushing and jostling for a glimpse of the latest wonder.

In 1762, George III's young wife, a German princess named Charlotte, received a zebra as a wedding gift. (We met Charlotte before, resplendent in a ball gown in a seaweed pattern.) The zebra immediately became a celebrity, and crowds scrambled to watch it graze near Buckingham Palace. It was "pestered with visits," one observer wrote, "and had all her hours employed from morning to night in satisfying the curiosity of the public."

Satirists churned out bawdy songs (but with discreet dashes) in honor of "the queen's ass." "A sight such as this surely never was seen / Who the Deuce would not gaze at the A - - of a Q - - - n?"

In 1767 two moose reached England, a gift from the governor of Quebec to the Duke of Richmond. The unlikely animals were "the most beautiful creatures ever seen in this Kingdom," one newspaper declared excitedly.

If a creature was strange enough, it could shove all other news aside. In London, on a November afternoon in the momentous year 1776, England's great and good craned their necks to see the animal in the center of an exhibition hall. Then they nervously clutched hands and braced themselves for an electric jolt.

They were seventy-one people altogether. The Duke of Devonshire was there, and the Marquess of Rockingham, and the president of the Royal Society with some thirty members of the society, and many members of Parliament, and also, according to the London newspapers, "other Gentlemen of Note, and Lovers and Encouragers of Science."

These were impressive pedigrees, but the star of the show was a four-foot-long electric eel, fresh from South America and one of the first ever to reach England alive. The company joined hands; two brave souls

* Not all royals were thrilled with living gifts. One official in Henry VIII's court drew the difficult assignment of writing a note to a noblewoman who had given Anne Boleyn a monkey as a present. "Of a truth, madam, the queen loveth no such beasts nor can scarce abide the sight of them."

clutched the eel, one grasping its head and one its tail; and "all felt its Electrical Stroke in the same Instant of Time." By good fortune no one died.

After the eel came a baboon in 1783 ("the Eastern Wonder or Child of the Sun") and a kangaroo in 1791 ("to be seen alive at the Lyceum . . . the only one in the Kingdom").

Similar wonders turned up well into the 1800s. It wasn't until 1847, for instance, that Europeans first learned of the existence of the biggest, most intimidating primate of them all, a beast called a gorilla.

For the scientists at the center of our story, struggling to sort out the world, these stories of endless discovery didn't *prove* that prehistoric animals still existed somewhere on the globe. But they kept that door enticingly open.

The centuries-long parade of unlikely animals had two contrary effects. On the one hand, the endless round of discoveries slowed science down, by raising false hopes that extinction might not be real after all. On the other hand, the spectacle of all those exotic creatures sped things up, by undermining religious dogma and freeing scientists to venture beyond old boundaries.

Everything turned on the question of how science and religion fit together. Until deep into the nineteenth century, as we have seen, nearly all England's leading scientists were devout Christians. This did not mean that they read the Bible literally. That had been a tough sell ever since Copernicus and Galileo had demoted Earth from its place of honor at the center of the solar system and set it careening around the sun.

Since *before* Copernicus, in fact, because many famous passages in the Bible required literalists to tie themselves in knots. They had devised ingenious theories to explain how it was, for instance, that God had declared, "Let there be light!" on the first day but had waited until the fourth day to create the sun and the stars.

Even so, the Bible remained *the* unimpeachable authority. The challenge was to interpret it properly, since, in Galileo's words, it was "often very abstruse, and may say things which are quite different from what its bare words signify."

The case of Noah and the ark was especially important. One of the best-known and most influential of all the Bible tales, the "bare words" of Noah's story inspired angry debates that took centuries to resolve.

Today, it is easy to forget how fierce the fighting was. Scientists in earlier generations had backed away from other biblical passages that didn't fit with new discoveries (Joshua had commanded the sun to stand still), but this was different.

Noah and the other stories in Genesis spoke directly about the everyday world of plants and birds and men and women, not about such arcane matters as the structure of the solar system. Jonah's story could be explained away with some fancy hand-waving; Noah's story was literally down-to-earth and far harder to skip past.

In the same years when modern science was taking its first steps, from about 1500 until late in the 1700s, brilliant thinkers continued to spend endless hours grappling with Noah's story. Scholars calculated the precise dimensions of the ark, and how many stalls it could fit, and how much space had to be set aside for hay. They fought over whether snakes needed special accommodations or could just wrap themselves around the rafters and whether fish could be left to fend for themselves.

We can dismiss those labors, but if we do we lose any chance of seeing the world as our forebears did. *What were they thinking?* we ask impatiently, forgetting that our descendants will look back at us and ask the same bewildered question.

Try for a moment to picture the world as our devout scientific forebears saw it. For them to put the Bible to one side would have required a mental revolution, perhaps as if sailors in that wooden-ship era had chosen to throw their maps and charts overboard and set sail, blind, across the ocean.

*   *   *

In ancient times, when comparatively few animals were known, the story of the ark may have seemed more plausible. A list of the animals cited in the Bible is long but far from endless—lions, cattle, deer, goats, ants, snakes, doves, and not a great many more.

With so few candidates, it may be just possible to picture how Noah could have crammed two of every species into the ark (even bearing in mind that God had ordered him to allow room for spares, so that he could offer sacrifices at journey's end).

But as time passed, the Bible story grew ever less likely. From around 1500 on, every voyage of exploration brought word of previously unknown species. By the mid-1700s, the great naturalist and classifier Carl Linnaeus had tallied some fourteen thousand species. The ark would have sunk under their weight.

(Linnaeus, a giant in the history of science, suffered a fate as cruel as anything in Greek mythology. He, who had assigned names to countless plants and animals, suffered a series of strokes in his old age and could no longer recall any of the names he had pinned to the world. In the end even his own name eluded him.)

One response from the leading thinkers of this era to the ever-growing catalog of animals was to bypass the hardest questions. Instead, they shifted to more congenial topics, as political candidates do in presidential debates. Walter Raleigh showed the way. He chose a small figure for the number of species on the ark, essentially by plucking a number out of the air, and then focused on room assignments.

"All these two hundred and eighty [species of] Beasts might be kept in one story or room of the Ark, in their several cabins," he wrote in 1614, in his *History of the World*, "their meat in a second; the Birds and their provision in a third, with place to spare for Noah and his Family and all their necessaries."

Raleigh was one of the great figures of his age—a poet, a soldier, a scholar, and a favorite of Queen Elizabeth (although he didn't throw his cloak into a mud puddle, as in the famous anecdote, so that she could keep her shoes dry)—and his lengthy examination of the ark story highlights its importance in the intellectual debates of his day.

*   *   *

But questions to do with New World animals could not be dodged forever. If Noah had brought animals of *every* species aboard the ark, why was there no mention of llamas or sloths or porcupines or any of their brethren? Where does Genesis talk of kangaroos hopping onto the cabin roof or possums playing dead on the deck?

And if God had indeed created such creatures in the beginning, how had they made their way to the ark in the first place, and then how had they found their way back home, after the ark finally came to rest on Mount Ararat?

Scholars banged their fists against their heads in frustration. "Why so many thousand strange birds and beasts proper to America alone?" wrote Robert Burton, the author of *The Anatomy of Melancholy*, in 1620. ". . . Were they created in six days or ever in Noah's ark?"

This was a sincere display of bafflement, not a bit of seventeenth-century snark. None of it made sense. How could it be, Burton asked, that these newfound animals were as different from those in Europe "as an egg and a chestnut"?

Other writers shared that bewilderment. Why were the world's animals dispersed in such odd ways? Thomas Browne, a seventeenth-century writer much admired for the elegance of his prose, pondered the riddle but confessed that he could not answer it. "How America abounded with Beasts of prey and noxious Animals," he wrote, "yet contained not in it that necessary Creature, a Horse, is very strange."

Not every writer was so cautious. Some proposed ingenious solutions that cleared up several mysteries simultaneously. A Flemish scholar named Justus Lipsius, writing around 1600, explained how rattlesnakes and bears had made their way to the New World (since no travelers would have carried them in their luggage).

The key was the lost continent of Atlantis, whose whereabouts Lipsius now pinpointed—Atlantis had once connected Africa to the New World, and animals had walked their way from Mount Ararat to their new home without ever getting their paws wet.

*   *   *

These were the considered views of important thinkers, not the fringe views of fanatics. But faith in Noah's story eroded away as new questions arose, old questions persisted, and contradictions surfaced. By the closing decades of the 1700s, scientists responded with ridicule to a text that their predecessors had read with reverence.

Think the story through, a German zoologist wrote in 1777. What would happen as soon as the animals left the ark? The two lions would devour the two sheep, and then they would proceed to gulp down the two cows, the two goats, the two buffalo. Soon, with all their prey gone, the lions would starve to death.

By the 1800s, geologists and paleontologists had taken the lead in arguing that the Bible should be read metaphorically. The Bible *could* have described a world in line with scientific teachings, wrote Hugh Miller, one of the most prominent Victorian geologists and a deeply religious man, but ordinary people would have been bewildered by such lessons.

Wisely, God had chosen a different course. Imagine the response, Miller told a lecture audience in the 1850s, if Genesis had described "a period in which there were lizards bulkier than elephants, reptilian whales furnished with necks slim and long as the bodies of great snakes, and flying dragons whose spread of wing greatly more than doubled that of the largest bird?"

Perhaps, in their horror and confusion, people would have abandoned religion altogether.

Miller and many of the other scientists we have met remained as devout as ever, but when it came to their professional lives, they ventured to move past the Bible. It offered guidance on ethical questions, they agreed, but not on scientific ones. To read the Bible literally, as their great scientific predecessors had done, was a mistake.

This was a momentous change, one of the great course shifts in Western history.

"A Cold Wind out of a Dark Cellar"

By the 1800s, when ichthyosaurs and plesiosaurs and their kin began turning up, scientists had stopped tormenting themselves with questions about Noah's ark. No one proposed that pterodactyls had marched aboard, two by two.

But to brush aside talk of the ark was not to brush aside the possibility that the world harbored unseen creatures. Thomas Jefferson, sitting in the White House in 1804 and hoping for word of mammoths and lions, was not a lone dreamer. As late as several decades into the 1800s, there were still holdouts who awaited news not just of bygone mammals but of living dinosaurs.

Some of them could cite impressive credentials. In 1819, for example, an English geologist named George Young suggested that the waving palm trees of a remote jungle might conceal huge, lurking creatures from prehistoric times. Or perhaps the oceans still contained monsters like Mary Anning's toothy ichthyosaurs and long-necked plesiosaurs, cruising the sunless seas.

Young was a Presbyterian minister who took for granted that the Bible was divinely inspired, but he was not a crank. A serious scientist and a well-rounded scholar, he was a linguist and a mathematician as well as a geologist. (He spoke Hebrew, Greek, Latin, French, and Italian, and dabbled in Arabic, Chaldean, and Syriac.) More important, he himself had discovered an ichthyosaur.

At a winter meeting of a natural history society, in 1819, Young gave a talk describing "a very curious fossil" that he had found a few months before in a seaside cliff in Yorkshire, in the north of England.

It looked a bit like a crocodile, in Young's account, somewhat like a dolphin, more or less like a fish. Young published several careful drawings that enabled scientists, many years later, to identify the particular species of ichthyosaur he had found. (There turn out to be many species of ichthyosaurs. Young's find was a different species than the one Mary Anning had found several years before.)

"This fossil animal" didn't resemble any living creature, Young conceded, but he encouraged his audience to keep their eyes open. "As the science of Natural History enlarges its bounds, some animal of the same genus may be discovered in some parts of the world."

There were, after all, plenty of hiding places out there. "When the seas and large rivers of our globe shall have been more fully explored," Young wrote, "many animals may be brought to the notice of the naturalist which at present are known only in the state of fossils."

We know now that even the most zealous search would have left Young disappointed. "Ichthyosaurs have been extinct for 93 million years," the historian Richard Ellis noted recently, which meant that any naturalists in quest of a living specimen were in for a long, frustrating hunt.

Young's view put him in a distinct minority, but he did have prominent allies. Thomas Thompson, an essayist who was a member of Parliament as well as a fellow of the Royal Society, contended that both William Buckland's megalosaurus and Gideon Mantell's iguanodon had survived into biblical times.

That would have brought them near the present day, though not all the way here. "I trust I have shown you," Thompson wrote in 1835, "that there is good ground for supposing that the leviathan of the Scriptures is the same animal as the now fossil megalosaurus; and that the behemoth was identical with the iguanodon."

Edward Nares, who was professor of modern history at Oxford for nearly three decades, endorsed the same view.* Perhaps the next voyage of exploration would return with a supposedly extinct beast chained to the deck and snarling at its captors. Other big names of the day chimed in along similar lines.

Ichthyosaurs and plesiosaurs like those Mary Anning had found might still "have their station in the subterranean waters," William Kirby wrote in 1835, "which should sufficiently account for their never having been seen except in a fossil state." Kirby was a serious figure, a fellow of the Royal Society who was best known as the author of the four-volume *Introduction to Entomology* (even by nineteenth-century standards, when extended courtships were common, this counted as a long introduction).

Some writers moved beyond speculation. Eyewitnesses reported sightings of possibly prehistoric escapees, and scientists eagerly debated the reports. In 1820, the *American Journal of Science and Arts* carried sworn testimony from travelers who had seen a sea serpent. "All this evidence, I think cannot fail to establish the fact that a large *Sea Serpent has been seen in and near the Bay of Penobscot*," one writer declared emphatically, and another *American Journal* correspondent filled in the picture of the creature. "His head was rather larger than that of a horse, but formed like that of a serpent. His body was judged more than sixty feet in length."

On the basis of many such reports, two eminent geologists agreed, in 1835, that "the ichthyosaurus, or some species of a similar genus, is still existing in the present seas."†

* Nares was a diligent scholar but a deadly bore. The historian Thomas Macaulay slogged through Nares's three-volume biography of William Cecil, chief advisor to Elizabeth I. "Compared with the labor of reading through these volumes," Macaulay wrote, with a characteristic mix of wit and complacency, "all other labour [such as] the labour of thieves on the treadmill" was "an agreeable recreation."

† Twentieth-century writers have occasionally suggested that the most famous sea serpent of them all, Nessie, is a plesiosaur whose forebears patrolled Loch Ness long before the birth of humanity. If only. It can't be true, because Loch Ness itself is only ten thousand years old.

*   *   *

For scientists to make a serious argument that even dinosaurs were not extinct shows how fraught the issue of extinction remained, well into the 1800s. To concede the possibility that a species could go extinct was to suggest that God's creation was flawed. But how could a perfect God produce imperfect work?

That baffling riddle set extinction apart from ordinary scientific questions—*Why do the stars shine? Why is Earth round?*—and made it far deeper and more troubling. The disappearance of a species would be not merely a crisis, the naturalist John Ray wrote, but something akin to a cosmic calamity, a "dismemberment of the universe."

Religion played a large role here, but so did everyday psychology. "The hint of extinction in the geological past was like a cold wind out of a dark cellar," wrote the paleontologist and essayist Loren Eiseley. "It chilled men's souls. . . . It brought suspicions as to the nature of the cozy best-of-all-possible worlds which had been created specifically for men."

Viewed solely as an idea, extinction might seem simple enough. But nothing to do with death could truly be simple. "Death was the first mystery," a French historian wrote in the mid-1800s, in a book on the birth of religion, "and it placed man on the track of other mysteries. It raised his thoughts from the visible to the invisible, from the transitory to the eternal, from the human to the divine."

All those *Why are we here?* questions arose when humans pondered individual deaths. Extinction brought unwelcome thoughts of a great scythe that took down whole quadrants of the world in a mighty swing. Was it any wonder that the world shivered?

Perhaps the greatest hazard was that coming to terms with extinction meant rejecting the belief that the world had a natural, God-given order. That belief ran so deep that it scarcely seemed like a belief at all. "For 2,000 years, there was an intuitive, elegant, compelling picture of how the world worked," writes the scientist and historian of science Alison Gopnik.

"It was called 'the ladder of nature.' In the canonical version, God was at the top, followed by angels, who were followed by humans. Then came the animals, starting with noble wild beasts and descending to domestic animals and insects. Human animals followed the scheme, too. Women ranked lower than men, and children were beneath them. The ladder of nature was a scientific picture, but it was also a moral and political one. It was only natural that creatures higher up would have dominion over those lower down."

This was conventional wisdom, and as usual, Alexander Pope expressed conventional wisdom better and more succinctly than anyone else.

> From nature's chain whatever link you strike,
> Tenth or ten thousandth, breaks the chain alike.

Worse yet, a single break would propagate instantly. Pope depicted the universal chaos that would follow the moment any link gave way:

> Let earth unbalanced from her orbit fly,
> Planets and suns run lawless through the sky;
> Let ruling angels from their spheres be hurled,
> Being on being wrecked, and world on world.

Ray issued his warning about the dismemberment of the universe in 1703, at the beginning of the century. Pope's poem dates from 1733, a few decades later. At the very end of the century, in 1799, Thomas Jefferson still held the identical view. No species could ever go extinct, Jefferson insisted, "for if one link in nature's chain might be lost, another and another might be lost, till this whole system of things should vanish by piecemeal."

It's important to note that Jefferson's rejection of extinction had philosophical roots, not religious ones. Jefferson believed in God but not in a God who reached into the contemporary world, and he had no patience for those who read the Bible literally.

He took the trouble to prepare his own copy of the Bible (which is now at the Smithsonian) by razoring out most of the passages that dealt with miracles and the supernatural, "or shall I say at once, of Nonsense." That left him with Jesus's moral teachings (and Noah's ark and heaven and hell), which stood out from the rejected material "as diamonds in a dunghill."

In summary, then, if you were a Christian who shared Jefferson's picture of a rational, mathematically minded God, extinction did not make sense, because nothing could be more offensive to reason than arbitrary gaps in creation.

And if you were a more conventional Christian, extinction did not make sense, because it implied that God did his best but occasionally had to issue recalls of faulty products.

But, though much of the scientific world had not yet heard the news, the debate was all but over. The extinction deniers did not know it, but by 1800 their fate was already as settled as if an asteroid had them in its sights.

CHAPTER 30

Sherlock Holmes Ponders a Bone

The asteroid came by way of France. On an April day in 1796, a brash newcomer to Paris stood up to present a scientific paper to the National Institute of Sciences and Arts. Georges Cuvier was only twenty-six, and this would be his first public lecture.

We have encountered Cuvier already, briefly, as the great sage handing down verdicts on the significance of fossils found by Mary Anning and Gideon Mantell and their peers. In 1796, Cuvier had already shown promise, but no one could have foreseen just how high he would rise.

In keeping with the new calendar put in place after the French Revolution, Cuvier delivered his lecture on 15 Germinal, Year IV (rather than the old-style April 4, 1796*). The revolutionary date suited a revolutionary message.

Cuvier skipped any sort of introductory remarks and began at once with what might have seemed a dry, technical observation: "Considerable differences have long been noted between the elephants of Asia and

---

* The calendar was short-lived, and so was a new way of counting the hours in a day. The new decimal system, which required new clocks and watches, called for ten hours in a day, one hundred minutes in an hour, and one hundred seconds in a minute.

those of Africa." Within a minute or two, though, it became clear that Cuvier was after big game, bigger even than elephants.

His goal was to demonstrate, swiftly and compactly, that extinction was a fact. What philosophers had declared impossible, what believers had called heresy, was true. *Listen for a moment*, he said, *and I'll prove it.*

Cuvier was a young man in a hurry (he was the youngest member of the National Institute), and Paris shortly after the revolution made an ideal home for him. The scientific institutions of the old regime had been replaced by newer versions—France's National Institute dated only from 1795 and its National Museum of Natural History from 1793—and the new organizations had searched out new talent. Cuvier, who had intelligence and ambition to spare but came from a family with no money or pedigree, quickly found his way.

He'd studied natural history and geology and, a year before his talk, had managed to snag a junior position at the National Museum. This was invaluable, because it provided him ready access to the best natural history collection in the world.

And as Napoleon steamrolled his way across Europe, looting as he went, that vast collection grew vaster still. A team of French scholars followed the army, scooping up books and plants and artwork and fossils "by right of conquest" and sending them to Paris in convoys of heavily laden wagons.

Over the course of Cuvier's career, the museum's collection would grow fourfold. He would work there for the rest of his life and live there, too, in a house on the grounds.

Cuvier had just begun settling into his new role when the museum received a natural history collection confiscated from Holland's deposed ruler. Two prizes from that collection—magnificent elephant skulls, one Asian and one African—drew his fascinated eye.

Cuvier was a brilliant anatomist who made a specialty of comparing bones and fossils from an enormous variety of animals. Crucially, the museum's ever-growing collection left him ideally positioned to compare fossil bones from ancient animals with bones from living species.

Tiny differences could signal to him that animals that looked alike were in fact separate species. Unexpected matches could prove, on the other hand, that the most unlikely pairs were in fact close relatives.

In time Cuvier would grow so famous for his deductive skills that in one of the Sherlock Holmes stories, Conan Doyle has Holmes himself profess his admiration. Just as the ideal detective could see an entire story in a single clue, Holmes says, "Cuvier could correctly describe a whole animal by the contemplation of a single bone."

He did it by focusing on bones and logic and "animal machines." Every animal was an intricate, beautifully functioning machine with a myriad of interacting parts, Cuvier explained, and a skilled investigator could deduce broad patterns from narrow clues.

If an animal happened to be a meat eater, he wrote, then "it is also requisite that the jaws should be so constructed as to fit them for devouring prey; the claws must be constructed for seizing and tearing it to pieces; the teeth for cutting and dividing the flesh; the entire system of the limbs, or organs of motion, for pursuing and overtaking it; and the organs of sense, for discovering it at a distance."

But it was the links between those perfect parts, and not the parts themselves, that truly sparked Cuvier's enthusiasm. As he explained his thinking, his cool façade gave way to something very near exuberance. Each insight sparked another in a *pop! pop! pop!* sequence, like firecrackers on a string. "To enable the animal to carry off its prey when seized, a corresponding force is requisite in the muscles which elevate the head; and this necessarily gives rise to a determinate form of the vertebrae to which these muscles are attached, and of the occiput [the back of the skull] into which they are inserted."

Conan Doyle had it right. Cuvier's approach was indeed akin to Sherlock Holmes's deducing a man's occupation from a callus on his thumb or a shiny patch on his cuff.

But many observers felt that praise along those lines *understated* what Cuvier and his fellow paleontologists had achieved. Putting flesh on bones was not just brilliant problem-solving but something more akin to a biblical miracle.

"Like the prophet Ezekiel in the vision," one paleontologist wrote exuberantly in the 1830s, "[we can] proceed into the valley of death, where the graves open before us and render forth their contents; the dry and fragmented bones run together, each bone to his bone; the sinews are laid over, the flesh is brought on, the skin covers all, and the past existence . . . starts again into being."

Cuvier's elephant talk was short, only four or five pages, but it displayed his gifts even so. Almost as soon as he began, he unveiled his surprise.

Present-day elephants weren't the real focus. Cuvier's targets were the mammoths found in Siberia and the incognitum that had won so much notice in America. "A scrupulous examination" of their teeth and jaws revealed that the Siberian mammoth and the American incognitum were distinctly different from each other. More important, both were different from present-day elephants, and—here Cuvier made sure no one could miss the point—both those imposing creatures had vanished from the earth!

"What has become of these two enormous animals of which one no longer finds any [living] traces?" he asked, and then he rattled off the names of several more creatures—prehistoric bears, rhinoceroses, and crocodiles—whose fossils he had studied. "None has any living analogue."

These were shocking claims, and Cuvier was not done. "Why, lastly," he demanded, "does one find no petrified human bone?"

The evidence shouted out a startling conclusion, with implications far beyond technical questions of anatomy. "All these facts . . . seem to me to prove the existence of a world previous to ours, destroyed by some kind of catastrophe. But what was this primitive earth? What was this nature that was not subject to man's dominion? And what revolution was able to wipe it out, to the point of leaving no trace of it except some half-decomposed bones?"

*   *   *

Today the smoke over the intellectual battlefields has long since settled, and few facts could be better established than extinction. It is not only real; it is all but inevitable. Ninety-nine percent of all the species that have ever lived are extinct.

A typical species lives between one million years and ten million years. The history of life on Earth calls to mind a cityscape seen from afar, the science historian Carl Zimmer has suggested, with some lights turning on and some blinking off and the overall number staying more or less unchanged. But every once in a great while, a blackout hits, and whole swaths of the city go dark in an instant.

The picture is a hard one for backers of "intelligent design" and creationism. "It doesn't seem so intelligent," writes one evolutionary biologist, "to design millions of species that are destined to go extinct, and then replace them with other, similar species, most of which will also vanish."

CHAPTER 31

Bursting the Limits of Time

C uvier's elephant lecture was a coup, and he followed it up immediately with a second triumph. As we have seen, Thomas Jefferson
had excitedly reported the discovery of a colossal American lion that he
dubbed megalonyx.

That excitement was short-lived, and it was Cuvier who squelched the
fireworks with a painstaking demonstration that the supposed lion was
in fact merely an oversized, sleepy-eyed sloth. No one had ever dreamed
there could be such beasts (modern-day sloths and anteaters are closer
in size to dogs than to minivans).

Cuvier was just getting started. He took it as his mission to begin
filling "the world previous to ours," where animals had reigned and
humans had never been. He did it with a stream of discoveries that propelled him to fame. Better yet, many of the creatures he resurrected had
once roamed the very territory where Cuvier now unveiled his finds.

By good fortune—for Cuvier and, as it turned out, for science
generally—Paris happens to sit atop a vast deposit of gypsum, a sedimentary rock that forms beneath warm, shallow seas. Gypsum can easily be
crushed into powder and fashioned into an excellent building material.
Drywall, which is all but ubiquitous on modern-day construction sites,
is made from gypsum.

The gypsum under Paris formed millions of years ago, and by Cuvier's era miners had been digging into it for two thousand years. (Hence "plaster of Paris.") Over the centuries they had found a great many fossils. Cuvier's National Museum had acquired a vast but badly organized collection.

Cuvier set himself the task of grappling with those bones, to find some sort of order amidst the jumble. This was, at first, more horrifying than enticing. For once, Cuvier's confidence seemed to slip. His tone calls to mind a hiker who happens on a cabin in the woods and peeks in the door to find a murdered family lying in bloody pools in the kitchen. "I found myself as if placed in a charnel house," Cuvier wrote, "surrounded by mutilated fragments of many hundred skeletons of more than twenty kinds of animals, piled confusedly around me."

Almost at once, though, Cuvier regained his footing. He switched into professional mode, and the voice of the shell-shocked eyewitness gave way to the cool observations of a detective at home in his grim element. "The task assigned me was to restore them all to their original position. At the voice of Comparative Anatomy, every bone and fragment of bone resumed its place. I cannot find words to express the pleasure I experienced in seeing, when I discovered one character, how all the consequences which I predicted from it, were successively confirmed."

In a public lecture soon after, Cuvier went into a bit more detail about how he worked. This was a setting that played to his strengths. A forceful and captivating speaker, he knew how to woo an audience.

He concentrated on large mammals, first of all, which everyone liked. (One of Cuvier's rivals had doomed himself, in the public mind, by focusing on uncharismatic creatures like worms and snails.) Better yet, Cuvier had found his large mammals in the Paris gypsum, and that gave his talk a kind of hometown appeal.

There was more, as the historian Martin Rudwick points out. The Parisian bones had scarcely been studied, which meant that their story was new *and* that Cuvier could claim the strange creatures he pieced together as his own. Moreover, they had been buried deep in the gypsum, so that extracting and assembling them gave Cuvier room to show off his anatomical chops.

His particular gift was an almost uncanny ability to see how a dip in a bone *here* might fit with a swell in a bone *there*. And fitting bones together was just the start. "The bones being well-known, it would not be impossible to determine the forms of the muscles that were attached to them; for these forms necessarily depend on those of the bones and their ridges."

With his animals' bones and muscles in place, "it would be straight-forward to draw them covered by skin." Why stop there? "One could even, with a little more boldness, guess some of its habits; for the habits of any kind of animal depend on its organization, and if one knows the former one can deduce the latter."

Over the course of the decade or so following his Paris lectures, Cuvier continued adding animals to what he might have called his Menagerie of the Extinct. By 1812, the tally had reached several dozen. The roster included mammoths and mastodons and cave bears and enormous elk and outsized turtles and crocodiles and hippos and rhinos and hyenas.

The world gasped in awe. "Is not Cuvier the greatest poet of our age?" Balzac asked. "Our immortal naturalist has reconstructed whole worlds out of bleached bones. . . . He shouts: 'Look!' And suddenly the marbles are teeming with creatures, the dead come to life again, the world turns."

Cuvier himself preferred a different analogy. He was not a poet or a sorcerer, he wrote, but what today we would call an archaeologist (the word did not exist until 1824). Pompeii had been identified only five decades before, and Cuvier likened himself to the much-admired "anti-quaries" at work there.

He was "an antiquary of a new order," Cuvier explained, whose con-cern was nature. But his approach and his goals were familiar. His aim, like that of conventional antiquarians, was reconstructing a picture of a vanished world. The difference was that they focused on "monuments" and ruins. Cuvier's monuments were fossils.

Cuvier suggested, as well, a second and even grander analogy. Although astronomers were confined to Earth, they had nonetheless "burst the limits of space." They had managed to deduce the motions of distant worlds that they would never see directly.

"Would it not also be glorious for man to burst the limits of time," Cuvier asked, "and, by a few observations, to ascertain the history of this world, and the series of events which preceded the birth of the human race?"

Though we pass our lives within a bubble whose extent in time and space is tiny, Cuvier declared, we can survey creation in all its vastness.

Cuvier, whose talent was rivaled only by his self-regard, was not yet the intimidating figure he would become, but he had already found the path that would speed him to scientific renown.

From early on, he made it clear that he was not a good man to challenge. Other scientists, he groused, indulged in airy speculation. Cuvier scorned their softheaded ways. Suspicious of religion, disdainful of theory spinning, trusting no tools beyond a sharp scalpel and his own sharp wits, Cuvier vowed to restrict himself to "modest anatomy" and "detailed study" and "scrupulous comparison."

His domain was the dissecting table, where theories could be put to the test, not the study, where imagination ran free. "It is not for us"— here the young rebel invoked the royal *us*—"to involve ourselves in the vast field of conjectures." *Here are the facts. Here are my conclusions.*

Dickens might almost have had Cuvier in mind when he invented Thomas Gradgrind, the school superintendent in *Hard Times*, who proudly described himself as "a man of realities. A man of facts and calculations. A man who proceeds upon the principle that two and two are four, and nothing over, and who is not to be talked into allowing for anything over."

Then came trouble, just when Cuvier was riding high. Despite his prodigious gifts—Cuvier was, in the judgment of one modern historian, "perhaps the finest intellect in nineteenth century science"—he never saw it coming.

Boiling Seas and Exploding Mountains

The problem Cuvier ran into was that the detective-style arguments that won him fame were, in a sense, *too* powerful. He had dazzled the world by showing that with just a glimpse at an animal, he could deduce its full portrait. Now he pointed out a second moral concealed in the same story.

The reason that tiny clues revealed whole sagas, Cuvier explained, was that every part of an animal had connections leading to every other part. And since those animals were not merely machines but perfect machines, a change of any sort would be a change for the worse.

But change was impossible, in any case, because refashioning one part of an intricate machine with countless interacting parts would require refashioning *every* part. And since animals were living machines, those changes would have to be made while the machine was running.

Cuvier set down his "everything everywhere all at once" principle as an axiom. "None of the separate parts can change their forms without out a corresponding change on the other parts of the same animal," he declared magisterially in his *Essay on the Theory of the Earth*.

That simple statement had profound consequences, and Cuvier boldly spelled them out. If animals could never change, you would expect the animals we see today to be identical to those that lived long ago.

They're not. And it was Cuvier himself who had proved it! Elephants, he showed, were different from mammoths and mastodons. Eagles were different from flying reptiles like pterodactyls, which Cuvier had been the first to identify.

Which was it? Did species change or didn't they?

Cuvier found an ingenious answer. It was, in a way, an intellectual coup as audacious as conjuring up an entire animal from a claw and a tooth. This time Cuvier looked at the fossil record and conjured up the history of Earth.

Species did *not* change, Cuvier insisted, even though fossils proved beyond a doubt that today's species are different from yesterday's. How could that be?

The only possibility was that some global calamity had come along and exterminated an entire world's worth of animals. Each of those unfortunate creatures had been perfectly designed for its annihilated world, to no avail.

Then, somehow—Cuvier made no suggestions, in keeping with his "Just the facts, ma'am" credo—God had created a *new* set of creatures, each of them a perfect machine perfectly suited to the new, post-catastrophe landscape.

Each individual species was fixed and unchanging, but the living world as a whole *did* change, because one set of unchanging species gave way to another. Cuvier made his case formally and without fanfare. "Life, therefore, has been often disturbed on this earth by terrible events," he wrote blandly, but he had in mind worldwide catastrophes of a type that far outdid present-day disasters like earthquakes or volcanoes.

If Vesuvius erupted, that would be a calamity, but the damage would be confined to a single region. Cuvier envisioned Vesuvius-scale explosions all around the globe. "The march of nature is changed," he went on, "and none of the agents that she now employs were sufficient for the production of her ancient works."

This was an unforeseen twist—Cuvier, the lover of facts and the archenemy of speculation, insisted that the facts led inexorably to the conclusion that the world had undergone a series of catastrophes beyond human imagining.

Cuvier insisted that this apparent twist was nothing of the sort. Look at fossils and layers of rocks, he demanded, and what did you see? You saw gaps and whole missing eras, as if you were studying a book that vandals had shredded. *He* was the one clinging stubbornly to solid fact. It was his self-indulgent rivals who had filled in those gaps with guesswork and inference.

But if Cuvier was a spokesman for restraint, he was an odd choice. The historian Donald Worster spells out the chaos that Cuvier had in mind. "These convulsions took on terrific, awesome splendor," Worster writes. "Mountains exploded, seas boiled, and monstrous creatures were buried in rumbling avalanches."

It might sound as if Cuvier had almost anticipated the theory of evolution. He had, after all, depicted past worlds filled with animals far different from those we see today. And the longer ago they lived, the more different from today's animals they appeared. But Cuvier fiercely opposed anything that smacked of evolution and anyone who dared advocate the despised doctrine.

Species were fixed, once and for all, he insisted, like stamps in an album. How else could Linnaeus have managed, half a century before, to pin double-barreled labels (*Homo sapiens*, *Canis familiaris*) to the world's flora and fauna? If species changed appearance in the course of time, as a snowman changes into a puddle, Linnaeus would never have had a chance.

Cuvier's claim was not that all dogs look alike—anyone can tell a Great Dane from a Chihuahua—but that all breeds share some essential quality of "dogness." It was that underlying resemblance that endured.

This focus on "essences" dated back to Plato, who had talked about everyday objects as mere copies of "ideal" objects that exist in the mind's eye. All triangles shared essential properties, Plato argued, and it was not important that some triangles were drawn in ink and others in chalk. In much the same way, a species was a permanent, unchanging feature of the world. One species could no more change into another than a lump of gold could transform into a chunk of lead.

But even as daunting a figure as Cuvier could not shut down the species debate simply by asserting that he saw no mystery in it. Cuvier detested theory spinning, but the beauty and variety in the natural world seemed almost to invite questions. The world is chockablock with species, for one thing, and it was natural to ask why there were so many.

And not only why so many, but why so many variations on a theme. Why, to choose one notorious example, are there 350,000 species of beetles? The most OCD-afflicted collector would have cried "Enough!" after the first ten thousand.

Evolution was one attempt at an answer. To Cuvier's dismay, the idea (the French called it *transformisme*) had its proponents. Jean-Baptiste Lamarck, an older colleague of Cuvier at the National Museum and his hated rival, was the best-known.

Lamarck argued that animals reshaped themselves by years of striving. Life grew ever more complex as time passed. Geography, not God, played the crucial role.

Giraffes were the classic example. Confronted with a landscape where leaves were hard to reach, giraffes strained upward. That effort paid off in slightly longer necks, which were passed on to offspring who grew necks slightly longer still. In tiny increments over the course of vast spans of time, old forms gave way to new.

Lamarck provided similar explanations for ducks' webbed feet, cats' claws, and horses' hooves, and that was just for starters. Here was a theory that explained both the history of the natural world and the myriad examples of marvelously designed creatures, and nary a deity in sight.

Lamarck is all but forgotten today, except perhaps as someone to mock. He was wrong about giraffes, and his other examples were equally off-base. A giraffe could spend its life stretching and straining and it would make no difference to its offspring, just as a musician's son could not expect to inherit his father's knowledge of how to play the trumpet.

But Lamarck had come to one crucial, correct conclusion at a time when nearly everyone else was wandering lost. He had recognized that species *do* change. The changes are hard to see because they are slow and our lives are short.

Everyone understands that a three-year-old careening along on a tricycle might one day march off to an office in a skirt and blouse. There's no mystery there, because individuals grow up quickly. But species change too slowly for us to see directly—stare as we like, we will not catch a wolf transforming into a poodle—so we think of them as eternal.

We happen to be stuck inside a time scale that obscures our vision. Geologists are the pros at dealing with great swathes of time, but their skill is hard to acquire. If laypeople could see through geological lenses, John McPhee writes, "continents would crawl like amoebae, rivers would arrive and disappear like rainstreaks down an umbrella, lakes would go away like puddles after rain, and volcanoes would light the earth as if it were a garden of fireflies."

In principle none of this should be a puzzle—no one struggles to explain why we can't see the hour hand of a clock move (or even the minute hand). But when we turn our gaze to the living world, it grows harder to remember that how things look is not necessarily how things are. Our blinkered view seems the only one possible. A French writer named Bernard de Fontenelle put the point succinctly. "No rose can ever remember seeing a gardener die," he wrote in the late 1600s.

Cuvier went wrong because, in modern terms, he was comparing snapshots of animals taken a few hours apart and asking, "Where's the change?"

Lamarck and Cuvier lit into each other. Lamarck struck first, with a critique that infuriated his fact-loving, theory-despising rival. Cuvier was a daydreamer, Lamarck charged, a fantasist who preferred inventing stories to grappling with reality. "A universal upheaval, a catastrophe which necessarily regulates nothing, mixes up and disperses everything," he complained, "is a very convenient way to solve the problem for those naturalists who wish to explain everything and do not take the trouble to investigate the course followed by nature."

(Lamarck didn't believe in extinction, although he allowed for a handful of human-caused exceptions like the extermination of the dodo. A "missing" species was simply one that had been transformed beyond recognition.) Cuvier snarled back (and in a remarkable venue—a eulogy supposedly in Lamarck's honor). A theory of evolution might "amuse the imagination of a poet," Cuvier declared, ". . . but it cannot for a moment bear the examination of anyone who has dissected a hand . . . or even a feather."

Some misguided scientists—Cuvier, whose manners were impeccable, named no names—had "laboriously constructed vast edifices on imaginary foundations, resembling the enchanted palaces of our old romances." They built their castles in the air unrestrained by rules or laws, blithely "piling up thousands of centuries with the stroke of a pen."

Cuvier worked his way to a mocking paraphrase of Lamarck's beliefs. "Only time and circumstance are needed for the . . . polyp to end by transforming itself, gradually and indifferently, into a frog, a swan, or an elephant."

The irony is that this absurd doctrine is a fine summary of not only what Lamarck believed but also of what virtually every biologist in the world today heartily believes.

Mayflies and Human History

C uvier and Lamarck's clash over evolution was, in a sense, perfectly timed, because it was a confrontation about the nature of change, and change was the watchword of the nineteenth century. Before the Industrial Revolution, the historian G. M. Trevelyan observed, life in England had moved at the pace of a slow stream.

Then, said Trevelyan, the placid stream suddenly acquired the momentum of a waterfall, "creating and obliterating one form of social life after another." Clamorous cities engulfed sleepy towns. Gaslights replaced torches and oil lamps. Steam engines did away with water-wheels. Speeding trains roared past horses and buggies.

The frenzy of creation and obliteration transformed ideas and beliefs just as it transformed every other aspect of life. Struggling to keep up with the flood of scientific discoveries, writers and thinkers found themselves intrigued and disoriented in roughly equal measure.

The poet and translator Edward FitzGerald (destined to become famous as the translator of *The Rubaiyat of Omar Khayyam*) took an exploratory voyage into the kingdom of time and emerged with his head spinning. FitzGerald was vastly learned and steeped in literature in several languages. In 1845 he wrote a letter to a friend remarking that he had been reading Homer, but "science every day more and more unrolls a greater epic than the *Iliad*."

"A few fossil bones in clay and limestone have opened a greater vista back into Time than the Indian imagination ventured upon for its Gods," FitzGerald wrote, "and every day turns up something new."

The new picture of time filled "the human Soul with Wonder and Awe and Sadness," in FitzGerald's judgment, but for many of his peers the wonder was no match for the sadness.

What seemed so dismaying to these early thinkers was not the expanse of time itself. Some insightful thinkers had glimpsed the truth long before the nineteenth century. (And some cultures had always envisioned nearly endless stretches of time. Hindus believed that a single day in the life of Brahma, the creator god, lasted four billion earthly years. In ancient Egypt the memorial temple for the mightiest pharaoh of all, Ramses II, was called "the House of Millions of Years.")

The West lagged behind, but Edmond Halley had peered into the past as early as the 1700s. Britain is an island, which means that thoughts of the sea are never too far off, but apparently, in 1715, Halley was the first to ask a simple question—simple in phrasing, at least, but vast in significance: why is the ocean salty?

"Whether something is obvious may not be obvious," the literary critic Christopher Ricks once remarked. Countless people have swum in the sea. How many have tasted the tang of salt water and thought, *Aha! Earth is very old*?

Halley (of Halley's Comet fame) was a rare character, not only for his intellectual gifts but also for the charm of his personality. He got along with everyone, including the formidable Isaac Newton, perhaps the greatest scientist who ever lived and certainly one of the fiercest. When Newton's archenemy Robert Hooke began whispering that *he* had come up with the theory of universal gravitation, and that Newton had stolen the idea from him—the theory was by many accounts the greatest achievement in the history of science—it fell to Halley to break the news to Newton. This was a mission akin to entering a lion's den to clean scraps of meat from his teeth. Halley managed it.

Halley's salt argument was simply put. Rivers carry water to the sea. Water from a river tastes fresh because it contains only a tiny bit of salt and other minerals. The sea is salty because it sits in place like a vast bathtub, accumulating whatever pours into it.* And since there is a *lot* of salt in the sea, we can be sure that a *lot* of time has passed since this story began.

Halley had it right, and in fact righter than he could have guessed. The amount of salt in the oceans is almost unfathomable. If you took all the salt in the sea and spread it evenly over the land, that layer of salt would be five hundred feet thick.

That makes Earth incredibly old, although Halley did not have the data he needed to come up with a figure (he would have needed to know how salty the ocean was in his day and also at some date in the past).

Seven or eight decades after Halley, astronomers got in on the "our home is old" act. The first was William Herschel, who began studying the stars in the late 1700s, an era when nearly all astronomers saw their task as studying the movements of the planets. They dismissed the stars as a backdrop.

Herschel was an astonishingly versatile man who began his career as a musician and composer and only took up astronomy when he was in his thirties. He designed and built his own telescopes, and soon they were acclaimed as the best in the world.

In astronomy's early years, scanning the skies could be arduous work. Herschel spent countless nights trying to hold a twenty-foot-long telescope steady in the wind and clambering up and down a ladder to his observing platform. On cold nights the ink in his bottle froze, and Herschel could not record any observations.

---

* Matters are not quite this simple, since some salt can leave the ocean (by combining with minerals in the ocean floor, for instance, or by being carried off in the form of sea spray).

The poet Thomas Campbell met Herschel in 1813, when the astronomer was old and celebrated. Herschel was soft-spoken, but his personal modesty made an odd fit with the grandeur of his subject. His picture of the cosmos left Campbell reeling. Scientists had known since the late 1600s that light does not travel instantaneously. Now William Herschel explained to his visitor what that meant.

(The first attempts to measure the speed of light had failed dismally. Galileo had tried, in 1638, with a scheme that involved two men on hilltops a mile apart, signaling each other by torchlight. In an era when watches did not even have minute hands, let alone second hands, the plan was doomed from the outset. Galileo might as well have tried to time a hundred-yard dash with a calendar.)

"I have looked further into space than ever human did before me," Herschel told Thomas Campbell. "I have observed stars of which the light, it can be proved, must take two millions of years to reach this earth."

That span of time was unimaginable, but Herschel piled one head-spinning proposition upon another. The stars that emitted that light, he explained, might no longer exist!

The reason was straightforward. A message that takes a long time to reach its destination carries out-of-date news; a letter from London to New York that takes a week to arrive carries week-old news. But when a message comes from across the cosmos, that simple truth seems not so simple. Campbell was staggered. He felt "elevated and overcome," he wrote a friend, "as if I had been conversing with a supernatural intelligence."

For the great majority of people in the nineteenth century, what was bewildering and sad—and not in the least elevating or uplifting—was that so much time had passed with scarcely a trace left behind. "It will strike terror into your soul," wrote Balzac in 1831, "to see many millions of years, many thousands of races forgotten," with nothing of them left but "piles of ashes."

Worst of all, there was almost no sign that *humans* had played any role in this long, long story. Charles Lyell, perhaps the greatest of all nineteenth-century geologists, spelled out the harsh truth. Earth was so old, he wrote in 1827, that it made human history "shrink into insignificance." Humankind had flickered into existence and would flicker out again, Lyell wrote, "as ephemeral . . . as the insects which live but from the rising to the setting of the sun."

Lyell, who had trained as a lawyer and had a gift for persuasive argument, made a compelling case that Earth had been shaped by slow, gradual changes rather than sudden, catastrophic ones. He did not downplay the impact of earthquakes and volcanoes—he devoted a great many pages to cataloging various disasters—but he rejected the notion of a *global* calamity, like a flood that drowned the entire world. Even massive earthquakes and volcanoes caused localized, short-lived devastation; they did not transform the globe as a whole.

Erosion wore down mountains, grain by grain, over unimaginable stretches of time. Rivers shaped valleys. Rolling hills and gentle summits testified to long exposure to wear and weather, just as an old man's wrinkled skin and bent back testified to the toll of the passing years. The present was the key to the past—the forces that shaped the world today were the same as those that had shaped the world in ages past.

So Lyell argued, and to great effect. But it was easier to agree that cliffs might crumble than to make peace with Lyell's observation about humans as akin to mayflies.

That message had been lurking just out of view since geologists had begun trying to estimate the age of Earth. As scientific findings accumulated, the truth had grown harder and harder to ignore, like the squeaks and groans of an old-fashioned wooden roller coaster inching its way to the top of a giant hill.

Lyell, whose religious views were conventional but not fervent, had no difficulty with humanity's diminished role. For the great majority of his contemporaries, it was shocking news.

*   *   *

The Victorians were nature lovers, as we have seen, but they had focused their attention on details—the colors of beetles' shells, the designs of birds' nests, the shapes of trees' leaves—and steered clear of big questions about life and meaning and humankind's place, except to hail God for his handiwork. But as more fossils turned up and evidence of Earth's ancient past accumulated, doctrines that had once seemed satisfactory came to feel glib and inadequate.

A modern-day historian of science rattles off questions that the Victorians found newly urgent. "If, as the Bible claimed, this planet had been made as a habitation for humanity," asks Jim Endersby, "why had its creator taken so long to get the tenants in? And if God was such a great designer, why was almost everything he'd designed now extinct?"

Against a backdrop of such riddles, talk of a "happy world" felt strained and artificial. Victorians still went on about vibrant flowers and chirping birds and "all things wise and wonderful," but now the familiar message came to seem, if not quite wrong, then overstated in a protesting-too-much way.

CHAPTER 34

Scattered by Desert Winds

Alfred, Lord Tennyson, the poet laureate of the Victorian age, left us what is perhaps the clearest account of how nineteenth-century thinkers struggled to reconcile the soothing truths of religion with the frightening news from the world of science.

Tennyson responded to the mood of his era like a tuning fork. His most admired work, "In Memoriam AHH," is a book-length, anguished attempt to find meaning and purpose in a world where science had undermined old values. The spur for the poem was the death of a beloved friend. In his despair, Tennyson likened himself to "an infant crying in the night."

"To thousands, this is a sort of sacred book," one Victorian writer remarked not long after Tennyson published his masterpiece, "and it dwells in their hearts in a place quite by itself." (Queen Victoria herself copied out excerpts from "In Memoriam" to ease her grief after the death of her husband, Prince Albert.)

Poetry was prestigious and popular in the nineteenth century to a degree that is almost unthinkable today, and Tennyson was perhaps the most popular poet of his day. When he died, in 1892, he was buried in Westminster Abbey with other heroes of the nation. Eleven thousand people applied for tickets to his funeral.

He began writing "In Memoriam" in 1833 after his closest friend died unexpectedly, of a stroke. From that single death, Tennyson moved quickly to vast mysteries about suffering generally. His first thought was that "God and Nature" cared only about the fate of species (which Tennyson called "types"). Perhaps nature was indifferent to suffering on an individual scale?

> So careful of the type she seems,
> So careless of the single life;

A moment later, Tennyson thought about fossils and prehistoric skeletons and saw that he'd been too quick to console himself. What was the message of all those lives entombed in cliffs and quarries?

> "So careful of the type?" but no.
> From scarped cliff and quarried stone
> She cries, "A thousand types are gone:
> I care for nothing, all shall go."

Tennyson never abandoned his belief in God—almost no one in his era would have done so, or even imagined the possibility—but he did find himself asking new questions about God's role. In his poem's early sections he had depicted nature as cruel. Now he wondered if maybe he had been wrong: perhaps God was *indifferent* to the vast suffering in the world. That prospect was even harder to bear.

Religion had always served Christians as a guide and a source of meaning, a resource in the face of sorrows. Believers had a personal relationship with God. He might punish you for your sins (or reward your good deeds). He watched over you. He listened to prayers. Now, horrifyingly, bewilderingly, came a new possibility—God might be unconcerned.

Tennyson pondered, once again, the record of death and mayhem etched in the cliffs. He grew more disheartened still. For one steeped in the scientific literature of the day, like Tennyson, the images of prehistoric beasts in mortal combat were hard to shake.

Nature was neither benevolent nor happy but soaked in blood. In a phrase destined to become one of the century's most famous, Tennyson wrote of "Nature, red in tooth and claw."

This was not casual musing or bar-stool philosophizing but agonized grappling with deep and urgent questions. (Tennyson wrote and rewrote "In Memoriam" over the course of seventeen years, and it stretched to nearly three thousand lines.)

Life was not sacred, it seemed, and perhaps it was not even significant. Tennyson posed a final, agitated series of questions. Were humans destined for oblivion just like all the other fossils? After all our loving and suffering and striving for truth, would our reward be to end up "blown about the desert dust, / or sealed within the iron hills"?

Tennyson was hardly alone in finding himself unhappy and unmoored in the new world that science and the Industrial Revolution had brought into being. We already encountered John Ruskin, the best-known art critic of the era, cringing at the sound of the geologists' hammers and lamenting his inability to hang on to his religious beliefs.

"You speak of the flimsiness of your own Faith," Ruskin wrote to a friend in 1851. "Mine, which was never strong, is being beaten into mere gold leaf."

Ruskin pointed an accusing finger directly at geology. In one of the best-known poems of the era, Matthew Arnold reached a similar destination by a different route. Arnold wrote "Dover Beach" around 1850 (but only published it in 1867). The poem is an elegant wail of lament, an ode to gloom.

Arnold disliked and distrusted science for what he saw as its worship of facts and its neglect of values. The world had sped ahead during recent decades, he conceded. In every way that could be counted or measured—miles of railroad track, output of factories, number of scientific discoveries—the signs of the era were energy and progress.

But there had been loss, too, and it was the loss of something real but hard to name that preoccupied Arnold. In this noisy, clattering era—its soundtrack would be an industrial symphony of hissing engines and thundering wheels—Arnold identified a theme nearly drowned out by the din.

For Arnold and his intellectual allies, the focus on machines and power and technology suggested an unnerving possibility. The universe itself might be a kind of machine, a soulless, churning engine that ran according to its own hidden rules, without a special place reserved for humankind.

The clanking sound of machinery made it nearly impossible to attend a quieter sound:

> The Sea of Faith
> Was once, too, at the full, . , .
> But now I only hear
> Its melancholy, long, withdrawing roar.

The world might appear bright and beckoning, Arnold continued, but that was an illusion. In truth, the world

> Hath really neither joy, nor love, nor light,
> Nor certitude, nor peace, nor help for pain;
> And we are here as on a darkling plain
> Swept with confused alarms of struggle and flight,
> Where ignorant armies clash by night.

It had been less than fifty years since Wordsworth's "My heart leaps up when I behold / A rainbow in the sky."

The world had continued to spin, but it had shifted course. The natural world had always been a refuge, a place to turn for comfort and consolation. No longer.

With doubts like those spelled out by the likes of Tennyson and Arnold, Paley's "happy world" came to seem remote and out of reach. Where cheeriness had once prevailed, melancholy spiked with anxiety reigned instead. Nature had changed its nature.

Victorians "went on loving their gardens and city parks, where *they* were in control," the historian Donald Worster observes, "but they were determined to dispel any foolish notions about the innate goodness of natural forces at work on earth."

The change in mood was partly a reflection of the ever-growing body of scientific facts. Weak arguments—that dinosaurs might be in hiding rather than extinct, say, or that a single flood had produced the thick, varied layers of rock found around the world—grew harder and harder to sustain. In time, what had seemed open-minded came to seem desperate.

One last-ditch effort to fend off the advance of science demonstrates the predicament of old-school believers. In the mid-1800s Philip Gosse was one of Britain's most admired naturalists. A writer and illustrator, he churned out charming and immensely popular books on birds and butterflies and the creatures to be found in tide pools.

His obsessions ran deep. Few excitements could match a sighting in a meadow or the arrival of a package, in the mail, from a fellow collector. On the September day in 1849 when his son was born, Gosse's diary entry ran, "E. delivered of a son. Received green swallow from Jamaica."

His story is worth a look because it highlights how the battle over natural history played out in the opening decades of the 1800s. Until well into the century, science and religion had seen themselves as allies, united by their delight in the wonders of creation. Scientists and believers were not enemies; scientists *were* believers.

But as the pressure to make sense of new facts grew, individuals had to find their own way to reconcile their clashing beliefs. Philip Gosse, an authority on every facet of the natural world and a fundamentalist Christian besides, found himself trapped in a closing vise.

Gosse belonged to a sect called the Plymouth Brethren that had turned its back on sinful, modern times. (Even children's books were forbidden to the Brethren, on the grounds that they told lies. "Never, in all my early childhood, did anyone address to me the affecting preamble, 'Once upon a time,'" Gosse's son recalled. "I was told about missionaries but never about pirates; I was familiar with hummingbirds, but I had never heard of fairies.")

As a devout Christian, Gosse fervently believed that the world was young, as the Bible taught. But how to explain fossils, which were plainly old?

Gosse proposed an astonishing answer. Fossils only *looked* as if they were relics of former lives. That was an illusion. In truth, fossils were "skeletons of animals that never really existed."

God had seeded the world with ready-made fossils, and careless thinkers had jumped to the conclusion that the world was old. Fossils were new artifacts made to look old, like pre-distressed jeans, and geologists had misunderstood what they were seeing.

This was not trickery on God's part, Gosse argued, but diligence. God had created each species out of thin air, essentially by snapping his fingers. Then he had fashioned a "pre-history" for every living thing. *That* was why Adam had a navel, though he'd never had a mother. It was why the trees in Eden had tree rings though they had sprung to life full-grown.

Thrilled with what he had deduced, Gosse laid out his ideas in a thick book and waited for the praise to flow in. He titled his masterpiece *Omphalos* (Greek for "navel," in honor of Adam) and gave it the subtitle *An Attempt to Untie the Geological Knot.*

"Never was a book cast upon the waters with greater anticipations of success than was this curious, this obstinate, this fanatical volume," wrote Gosse's son Edmund. "My father lived in a fever of suspense, waiting for the tremendous issue. This *Omphalos* of his, he thought, was to bring all the turmoil of scientific speculation to a close, fling geology into the arms of Scripture, and make the lion eat grass with the lamb."

Alas, the book's only effect was to end Gosse's career. The findings of geology had grown too substantial to dismiss.

Gosse's pratfall brought him no allies. His attempt to wave away the evidence that Earth was old marked him as a man of the past. He had missed the boat. He had, in fact, dressed up in topcoat and tails, waved farewell to a crowd on the pier, stepped forward from the gangway, and then, to his own astonishment, found himself plummeting into the harbor depths.

Lizards in Scripture?

While intellectuals like Tennyson and Matthew Arnold flailed about and the public struggled, scientists themselves proceeded merrily along. This was a surprise, because they had been forced to abandon some of their deepest beliefs.

Naturalists and philosophers had long recoiled at the thought of extinction, for instance. The loss of even a single species, they had warned, would bring chaos. Then Cuvier came along preaching a sermon of *universal* death, and, rather than tar and feather him, thinkers responded by smothering him with praise.

Loren Eiseley, the historian of science, marveled at the shift. "From the idea that one lost link in the chain of life might cause the whole creation to vanish piecemeal, man had passed, in scarcely more than a generation, to the notion that the entire world was periodically swept clean of living things."

That was a stunning about-face. But English scientists barely acknowledged that they had changed course even a little. The key was finding a way to merge Cuvier's new scientific doctrines with their own long-established religious beliefs. Cuvier had proclaimed, after all, that the most indisputable fact in geology was that "the surface of our globe has undergone a great and sudden revolution, the date of which cannot be referred to a much earlier period than five or six thousand years ago."

He had in mind a vast deluge, Cuvier wrote, but he never stated outright that he was referring to the biblical flood. His English admirers happily filled in that gap. What could have been more tempting than to enlist one of the great names of the day in their ranks?

Though the English were quick to embrace him, Cuvier did not truly belong among the believers. He was far less devout than his English colleagues, and he believed that science and religion should be kept apart from each other, which was exactly contrary to the English view. The deluge he had in mind was a natural event, not a display of divine wrath.

But the geologist who translated Cuvier's opus into English, a Scot named Robert Jameson, depicted the flood and Cuvier's other "catastrophes" as supernatural events. English scientists gladly went along with that sleight of hand.

This was not hypocrisy. England's leading scientists were sincere believers, but they took for granted that the Bible needed interpretation if its true meaning was to emerge. Just as landscapes and cliffsides conveyed messages that geologists could read but laymen would not recognize, so did biblical texts contain messages that were easy to miss or misunderstand.

These seeming conflicts weren't seen as contradictions or oversights but as opportunities. *Plainly* the days of creation in the Bible, to take one notable example, were not days in the ordinary sense.

When an old man says "in my day," scientists noted, he means "a long time ago." In much the same way, biblical "days" surely represented eons. Nor did "in the beginning" refer to the beginning of the week. It referred to an unimaginably distant era in the past when the world had taken shape.

So England's leading scientists argued. The Bible was "addressed to the heart and understanding, in popular forms of speech," explained the geologist Adam Sedgwick. That was why passages in the Bible depicted God as "capable of jealousy, love, and anger." Those descriptions were not meant literally, Sedgwick wrote, and scripture was "not . . . nor [did] it pretend to be, a revelation of natural science."

*   *   *

Dinosaurs presented a particular challenge. The Bible never mentioned a word about them. Why not?

William Buckland, perhaps England's best-known geologist, tackled that one. Buckland's idea was that, although the world was ancient, the Bible picked up the story only recently, at the crucial moment when humans appeared on the scene. What had happened beforehand was of no religious significance.

"It stood to reason that there was no mention of extinct lizards, or saurians, in Scripture, for these were not self-conscious, moral beings, and played no part whatsoever in God's purpose as narrated in the Bible," explains the historian of science Allan Chapman. "This purpose was to establish and proclaim his creation of a world, post-Edenic, for those beings whom God had made in his own image: Adam and Eve and their descendants down to this day."

Hugh Miller, like Buckland both a devout believer and a geologist, took a different path to a similar conclusion. Yes, it was true that "saurians" were extinct, but this was a case where extinction was to be celebrated rather than feared. Here was still more proof that God had designed a happy world!

Miller focused on Earth's long history. The world had once been ruled by mega-reptiles, he noted, and then, much later, by mammals. Where the two were still found together—Miller cited a boa constrictor wrapping its "terrible coil" around a hapless deer—the story was apt to end poorly.

Fortunately, the Divine Planner had made careful plans. "A world which, after it had become a home of the higher herbivorous and more powerful carnivorous mammals, continued to retain the gigantic reptiles of its earlier ages, would be a world of horrid, exterminating war."

Rather than sanction such mayhem, God had wiped the "saurians" off the board. He had replaced those mighty reptiles with puny and harmless successors, mostly, the likes of lizards, turtles, skinks, and geckos. As a result of that wise decision, Miller noted contentedly, "life is, in the main, enjoyment."

*   *   *

From our vantage point, these cheery responses come as a surprise. It might have seemed that geology's new lessons would undermine religious faith. Instead, they bolstered it.

Vistas of endless time, for instance, often inspired awe. "The sheer scale and unanticipated strangeness of the earth's long history," writes the historian Martin Rudwick, "was often treated as welcome new evidence for the grandeur of God's creation."

So "saurians" were fine, and even the most devout scientists could ponder them (and their extinction) without qualms. Which was fortunate, because it would have taken a formidable bit of housekeeping to sweep ten-ton lizards under the rug. And then came another twist in the tale.

This one would put the spotlight squarely on ichthyosaurs and plesiosaurs and the rest of the prehistoric menagerie. All those beasts would soon be enlisted as a sort of supersized cavalry, where they would serve as troops in a battle against a dangerous and formidable foe.

The general in that battle was one of England's greatest and most controversial scientists, a strange, hugely talented man named Richard Owen. The foe was the theory of evolution, which (even in its pre-Darwinian form) threatened to shove God aside and upend the entire social order.

Dr. Jekyll and Mr. Hyde

Richard Owen was a complicated man. He was brilliant, backstabbing, charming, and manipulative. He worked relentlessly—he described *hundreds* of species new to science—and his knowledge was unsurpassed. So was his zeal for cultivating useful friends and trampling enemies. Over a long career, he rose to the top of the British scientific establishment. He hobnobbed with Queen Victoria and Prince Albert, and he accumulated too many honors to count (including a knighthood and, posthumously, a larger-than-life statue).

But Owen made people uneasy. Partly this had to do with his area of expertise. Owen spent his days staring at old bones and carving up newly deceased animals that had been stewing in formaldehyde. "As an anatomist," one historian remarks, "he carried the whiff of death."

Owen's appearance worked against him, too. Tall, thin, and bony, with an immense forehead and a jutting chin, he had the misfortune of looking a bit like Uriah Heep, one of Dickens's most memorable villains. Heep was so slimy, Dickens wrote, that when he perused a book, "his lank forefinger followed up every line as he read, and made clammy tracks along the page . . . like a snail."

Owen inspired the same kind of repugnance (and, according to Thomas Carlyle, did indeed have a "clammy, irresponsive hand"). His dark, deep-set eyes reminded nearly everyone of the huge, blank eyes of the prehistoric reptiles that he had made a career of studying. As Owen aged, he grew ever more cadaverous. In time it was not only the cave-deep eyes that drew attention. "As if by sympathetic magic," one biographer writes, "he began to resemble his vertebrate fossils."

It was Owen's manner, most of all, that put people off. "The truth is," Thomas Huxley observed, "he is the superior of most, and does not conceal that he knows it." Huxley was a rival of Owen's and a fierce enemy, but many others made similar observations.

Owen was, in fact, a cultivated man, a good singer who played the cello and flute, a lover of the theater, a friend of Dickens and Tennyson (and nearly everyone else who was anyone in Victorian England). But he was transparently ambitious, and he tended to fawn on the powerful and ignore the lowly. Jane Carlyle, the wittily malicious wife of

Thomas Carlyle, likened Owen's smile to "sugar of lead," a poison with the unusual property that it tasted sweet rather than bitter.[*]

Owen had studied to be a physician, but it had been the mysteries of anatomy, and not the treatment of patients, that drew him. His career began early. As a sixteen-year-old in the grip of what he called an "anatomical passion," he had apprenticed to a surgeon who was responsible for the care of inmates in the local jail. There Owen's first encounter with a dead man—he pulled back a sheet covering the face of a young prisoner who had died of a fever, to reveal "pale, cold features and glassy, staring eyeballs"—left him almost faint with terror.

Six weeks later, he encountered far worse. For the rest of his life he would retell the story of a January night in 1821. When he heard that a prisoner in the jail's hospital had died, he slipped a few coins into a guard's hand and told him that he would be back that evening to conduct some private business. He returned a few hours later carrying a large paper bag. Once inside the morgue, Owen closed a heavy wooden door behind him and, safe from interruption, sawed the dead man's head free from its body. He stuffed the head inside the paper bag.

Then he raced down the stairs and into the night, with the bag hidden under his cloak. Out on the street, Owen slipped on the ice and crashed to the ground. The head spilled out of the paper bag and bounced downhill.

Owen chased after it, skidding as he ran. At the foot of the hill the head smacked against a cottage door. Someone inside flung the door open and, at the same moment, Owen crashed through. He grabbed the head from the floor, shoved it under his cloak, ignored the shrieking from inside the cottage, and ran off.

Was it true? It seems unlikely, but why would Owen tell and retell a story that cast him in such a bad light?

---

[*] Jane Carlyle thought poorly of Owen, but she had a clever, dismissive word for nearly everyone. Thomas Carlyle was less witty but more irritable. "It was very good of God to let Carlyle and Mrs. Carlyle marry one another," one of their acquaintances remarked, "and so make only two people miserable and not four."

Owen's misadventure took place decades before Robert Louis Stevenson created Dr. Jekyll and Mr. Hyde. But even without that example to draw on, an acquaintance of Owen's called him "one of the oddest beings I ever came across" and described him as bewilderingly complex.

Owen seemed "as if he was constantly attended by two spiritual policemen," his friend wrote, "the one from the upper regions and the other from the lower, the one pulling him towards good impulses and the other towards evil."

The "odd being" would soon find himself at work in a setting—a museum with a collection out of a nightmare—that might have been dreamed up by Stevenson at his most fevered.

Defunct Animals and Open Windows

In 1825, Owen began practicing medicine in London and, what was more to his liking, took a job trying to impose some order on the gigantic, unruly collection of the natural history museum at the Royal College of Surgeons.

This was the Hunterian, the overstuffed attic of a museum we ventured into in chapter 15, when Gideon Mantell turned up looking for clues that would help him identify his iguanodon tooth. The museum had been founded by John Hunter, a renowned surgeon and anatomist and a hugely controversial figure in his day. His many allies saw him as a genius and an innovator far ahead of his hidebound peers. His rivals saw only a fanatic who spent his days obsessively cutting open anything he could get hold of, from earthworms and elephants to human corpses that grave robbers had procured for him.

By the time Owen arrived, Hunter had been dead for more than three decades, and the collection he had amassed had gone nearly untended all that time. Five thousand glass jars filled with floating embryos and malformed snakes and lizards lined endless shelves. Whale skeletons hung from the ceiling. A wall display featured the tattooed arms of South Sea islanders. A table showed a lineup of skulls from humans, chimps, dogs, and crocodiles.

The museum's star exhibit was a towering human skeleton that Hunter had obtained in a ludicrously shameless way. In the spring of 1782 all London had been abuzz with talk of the "Irish giant," an elegantly dressed, well-spoken young man who supposedly stood an amazing eight foot two inches tall. (His true height was seven foot seven.)

Charles Byrne had already performed—which is to say, stood onstage—in Scotland and northern England. Now London would have its turn. For a fee of about $15 in today's money, the public could gawp at, in the words of one excited news story, "a prodigy like none that had ever made its appearance among us before."

Byrne, who appeared under the stage name O'Brien, was a sensation. He met the king and queen, and the Royal Society arranged a private viewing for its members. Hunter, on the lookout for an exhibit that no museumgoer could resist, saw an opportunity.

Recognizing that the giant was in bad health, he proposed a deal: he would pay, now, for the right to display the giant's skeleton in his museum, later. Horrified, Byrne said no.

But Hunter's forecast proved accurate, and the next year, 1783, found Byrne on his deathbed. He spent his final hours making plans intended to keep him out of Hunter's clutches. The scheme was simple—Byrne arranged for some fishermen to take his body out to sea after he died, weigh it down, and send him into the deep and out of reach. Alas, Hunter found the fishermen and paid them to deliver the giant's corpse to him. Then he took it home and boiled it down, in a giant cauldron.

Soon after, Charles Byrne's skeleton took its place front and center in Hunter's museum, where it remained for two hundred years.[*]

Trying to organize the fourteen thousand specimens in the Hunterian Museum's motley collection was daunting work. Owen powered ahead with fanatic diligence, putting in long hours organizing,

---

[*] Byrne was only twenty-two at his death. His gigantism was due to a pituitary tumor, according to DNA studies carried out in 2011. The Hunterian Museum announced in 2023 that Byrne's skeleton would not be displayed again.

Queen Elizabeth II inspecting the Irish giant
at the Hunterian Museum in 1962

cataloging, and dissecting. He lived above the shop, in the museum building, which made it easier to work deep into the night.

At the same time, he'd taken on a host of anatomical projects that had nothing to do with the museum. One biographer tried gamely to convey the range of Owen's investigations. "He published prolifically... on the deceased inmates of the Zoological Gardens, the reproductive organs of marsupials and monotremes from the Australian colonies, on the great apes, dinosaurs, and the extinct flightless birds of New Zealand."

Along the way Owen managed to acquire a wife whose devotion to natural history nearly matched his own. Caroline Clift had grown up in the Hunterian Museum. Her father—and Owen's boss—was a brilliant, self-taught anatomist and the museum's curator. Caroline herself was intelligent, interested in everything, and unflappable. Nothing rattled her, whether it was her husband conducting a postmortem on a kangaroo or Charles Dickens turning up for dinner. (Caroline served apple tart for dessert and noted happily that "Dickens enjoyed it like a schoolboy.")

Caroline was a lively writer—Richard's prose style was, in one historian's judgment, "at best stupefyingly dull and, at worst, incomprehensible"—and in a journal that she kept for decades, Richard appears warm and sociable, utterly unlike the dark and fearsome creature his scientific rivals depicted. ("I determined I would never love any but a very superior man," she told the other guests at a dinner party half a dozen years after her marriage, "and see how fortunate I have been.")

But even seen through Caroline's forgiving eyes, Richard comes across as remarkably single-minded. Anatomy was his north star. He had contrived to have first dibs on any animals that died at the new zoo in London, for instance, and he found ways to obtain countless other creatures, too.

There were virtually no bounds to his curiosity. In Owen's day England heard of platypuses for the first time. With their webbed feet and duck bills, these strange hybrids didn't seem to fit in any ready-made category. Were they mammals? Reptiles? Hoaxes cobbled together by troublemakers? (Museum curators today pass along a story that the first platypus to reach London had marks on its bill from a pliers, from someone's attempt to yank off the bill and uncover a fraud.)

Owen opted for mammal. To test his theory, he needed to find milk. He took a platypus that had been floating in a jar of formaldehyde for a year, ever since its arrival from Australia. Now for a tasting! But platypuses have no nipples. Where could the milk be hiding?

Owen proceeded to knead the platypus's breast tissue. To his delight, "minute drops of a yellowish oil" soon appeared. Owen tasted the drops. They didn't taste or smell like milk (or like much of anything, he noted, except "the preserving liquor"). This was an inconclusive finding, Owen admitted, and he conceded that he could not quite mark the mystery as solved. (Owen had nearly cracked the riddle. Platypuses do feed their

This drawing of two platypuses comes from a three-volume work
called *The Mammals of Australia* by John Gould, which appeared
between 1845 and 1863 and provided many Europeans with their
first looks at Australia's unfamiliar animals

young on milk, which oozes out of ducts in their mammary glands and
collects in shallow grooves in their skin. Newborns slurp and suck drops
of milk from their mothers' skin.)

When he wasn't licking platypus fur, Owen spent endless hours at the
Hunterian Museum pondering bones and reconstructing skeletons. A
more or less typical entry from Caroline's journal, from January 29, 1838,
conveys the tenor of his days. "Today R. cut up the giraffe which died at
the Zoological Gardens. Afterwards he went to the Royal Institution to
dissect a snake. They have now got the skeleton of the hippopotamus
up in the museum."

Caroline recorded it all as if nothing could have been more routine.
When a rhinoceros died at the zoo one December day in 1838, Richard
ordered it sent to his home so that he could get to work without delay. "The
defunct rhinoceros arrived while R. was out," Caroline wrote in her journal.
"I told the men to take it right to the end of the passage where it now lies."

Defunct became one of Caroline's favorite words. A "defunct elephant" was even worse than the defunct rhino. "It made me keep all the windows open, especially as the weather is very warm," Caroline wrote in June 1847. "I got R. to smoke cigars all over the house."

Years of similarly varied and intense days left Owen without a challenger as the greatest expert since Cuvier on the anatomy of creatures both living and extinct.

From the start Caroline was a partner and ally, not simply a put-upon spouse. "I made two ink outlines of shark's teeth"—she was a skilled illustrator—"and tonight translated from the German for R.," she wrote on one winter evening. "After that I read aloud from Cuvier whilst R. compared the editions. Wrapped up the tortoise in flannels"—this was a pet who liked to be warm—"before I went to bed, and put it in the front cellar."

Another evening found Caroline back at her scientific illustrations (and Richard in an autocratic mood). "Engaged all day in drawing a wombat's brain for R. When R. came in he said it was all wrong, so I must do it all over again."

But life was not all work and wombats. "To luncheon at Dean Buckland's," Caroline wrote on March 22, 1847. The two couples were fond of each other, but meals chez Buckland were never routine because Buckland was perpetually serving up experimental dishes like hedgehog or mouse.

On this March day, the experiment seemed less challenging than usual. "A piece of roast ostrich, which we all tasted," Caroline wrote. "It was very much like a bit of coarse turkey."

The next day's diary entry was brief. "R. had a very bad night. Query, roast ostrich?"

The Mystery of the Moa

Dining mishaps aside, the 1840s were good years for the Owens. Until about a decade before, Owen's career had been stalled. He was assistant to Caroline's father, William Clift, but his pay was low, and the job prospects for anatomists were limited. More important, Clift had arranged long before that his son, William Home Clift, who worked with him, would someday take over his role as head of the museum.

The younger Clift (who was Caroline's brother) had been groomed for his position since childhood. At age sixteen, when he was already at work in the museum, his father had sent him instructions for his next task. William was "to try his best in cleaning and setting up the Oran Outang." Father and son were both talented and good-natured, but an anointed son far outranked an interloper, and Owen seemed doomed to the periphery.

Then, one September evening in 1832, young William Clift hailed a cab. The driver sped down Fleet Street, in London, and careened around the corner onto Chancery Lane, which was much smaller. The cab flipped. William flew into the air and smashed his head on the street. He was taken to St. Bartholomew's Hospital, unconscious. The doctor on duty was Richard Owen.

Nothing could be done. Clift's skull was fractured. He lingered a few days and then died, at age twenty-nine.

The death of his son was "naturally a great grief to Mr. Clift," Owen's grandson would later write in a biography of his famous relative, "but at the same time it was a consolation to know that Owen would eventually stand in the place of his son, both in the museum and at home."

And so he would. Almost at once, Owen took on new responsibilities and began pulling down a far larger salary. He had always had to scramble for money—Caroline's mother had refused to allow her daughter to marry a man with such poor prospects, and it had seemed their engagement might drag on forever—but now the picture brightened.

After an eight-year engagement, Richard and Caroline married in 1835, on Richard's thirty-first birthday.

Though Owen's gift for making enemies was as outsized as his other skills, no one denied that he possessed daunting scientific talent. In 1839 a sailor brought him a single fragment of bone from an unidentified creature in New Zealand. The bone was big, about six inches long, and Owen thought at first that it might have come from an ox.

But it was light in weight, with an open honeycomb structure, and markings on its outer surface looked like ones that Owen had seen on the surface of an ostrich's thigh bone (though the bones were different in shape).

Owen made painstaking comparisons of the new bone with bones from fourteen species, including humans, kangaroos, and a giant tortoise. Then he made a daring guess: the new bone came from an enormous bird even bigger than an ostrich, a giant bird much too heavy to fly.

This was particularly bold, Owen pointed out, because the bone fragment truly was bone, and not a fossil, which meant that it was not ancient. As far as anyone knew, the biggest bird in New Zealand was the size of a pheasant. Yet here was Owen suggesting that a bird twice as tall as a human being might be strutting around petite New Zealand, unnoticed.

Owen wrote up his theory for the Zoological Society of London and sent one hundred copies of his essay to New Zealand, asking for more information. (The Zoological Society left Owen on his own, refusing to print his paper unless he explicitly agreed that the responsibility for its dubious proposal "rested solely with the author.") For three long years, not a word came back from New Zealand.

Out on a conspicuous limb and "anxious" about looking silly, Owen paced and fidgeted. Finally, in 1843, a missionary in New Zealand sent him a giant box of bones.

An eyewitness had described the scene years before when Owen had first examined the mysterious bone fragment. "He took, in our presence, a piece of paper and drew the outline of what he conceived to be the complete bone."

Now the same observer was on hand again, eager to see how the story played out. Owen rummaged inside the missionary's box. "When a perfect bone arrived and was laid on the paper, it fitted exactly the outline which he had drawn."

Owen stands next to the giant skeleton of a moa. In his hand he
holds the single bone that held the key to the mystery.

Owen assembled the rest of the bones—here stood a twelve-foot-tall bird, never before seen, now called a moa. (It turned out that moas were extinct, though they seem to have survived until a few centuries ago.)

Prince Albert himself rushed to see the giant bird and to shake Owen's hand.

Richard's reputation rose steadily through the 1830s and '40s, and Caroline seemed to revel in the surprises that perpetually came her way. On a November evening in 1843, she wrote in her journal, there was a ring of the bell, "and in another minute there suddenly stalked in a magnificent, tall American Indian in full dress—paint, necklaces, and tomahawk, and a red mantle over all."

Even Caroline was taken aback for a moment. "I felt rather staggered but endeavoured to show no sign of it, and so asked the gentleman to sit down in the armchair, which he did in a calm, well-bred manner. He was accompanied by a young gentleman, a native of Guernsey, but who had lived some time among the Indians. We were very soon quite at ease with each other."

Richard decided to show their visitor his museum. He looked politely at various prized objects, but even the giant's skeleton roused only mild interest. "When he had seen O'Brien he made a remark which, being interpreted, was, 'This is large.'"

Once the museum tour was complete, the little group settled in to look at natural history books with plates of various animals. The books were a success. The "chief" chose a picture of a leopard as his favorite, Richard ordered up some wine, and hosts and guests passed a cheery evening flipping through the pages.

In the 1840s, Owen was at the peak of his powers and poised to fight the great intellectual battle of his era. This was the moment, he decided, to bring dinosaurs onstage.

"The Invention of Dinosaurs"

In the first half of the nineteenth century, evolution was the subject of endless confused debate. This controversial new theory purported to answer two questions, one new and one old.

The old question was how animals had come to be so well designed. How to explain a dolphin's grace or a lion's grandeur? The new question was why life seemed to have a direction, from primitive beginnings long ago to a complex flowering today. Why was it, people wanted to know, that in the oldest, deepest layers of rocks you found no fossils at all, or fossils of invertebrate creatures like snails and clams; and in layers not as old or as deep you found reptiles; and in still younger, shallower layers you found mammals?

The traditional answers to both questions were simple. God had arranged things that way. He had delighted in painting the peacock's tail, and he had crafted the ultimate Horatio Alger story, long before Horatio Alger was ever born. From a lowly starting point, when the world had featured little but bugs and worms, God had ordained that life would grow ever grander and more sophisticated until at last the time came for humankind to step into the spotlight.

Now evolution had come along and proposed new answers that, threateningly, seemed to push God to the sidelines. The new theories were vague and hard to understand, but the idea seemed to be that some "force" or "drive" or "impulse" shaped individual animals and also propelled the history of life uphill, along a path of ever-growing complexity.

Just how that mysterious force worked no one seemed able to explain, but evolutionists did agree on one crucial point. The force at the heart of the story was natural, not divine. Nature was a machine, they asserted, and in time science would explain its workings.

Evolution in this era was mostly associated with France. That was not a point in its favor, at least not in England. (This was evolution before Darwin—he would choose to remain safely out of view until 1859. Before then the public knew him only as the author of an exciting travel book called *The Voyage of the Beagle*, with eyewitness descriptions of volcanoes, earthquakes, and "savages.")

For decades after the French Revolution, Englishmen who heard the word *France* thought of atheism and unrest and rebellion. Even without that taint, evolution would have been unwelcome. But *with* that baggage, a theory that aimed at nothing less than reframing the story of the natural world stood little chance of a fair hearing.

Evolution was, after all, a theory built around change. Those changes had to do with plants and animals, not courts and laws and taxes and reforms. But the fear, among conservative thinkers, was that talk of change somewhere might spill over to talk of change elsewhere. If the world *could* be different, perhaps the world *should* be different. Far better to stick with a doctrine that endorsed a world of fixed, permanent roles, where everyone had a place and everyone knew his place.

Few imported doctrines could have been less welcome than evolution. In England from the 1830s to 1859, writes one modern paleontologist, "the suspicion of being a transmutationist [i.e., supporter of evolution] was every bit as damning as the suspicion of being a communist during the McCarthy era in the United States."

English scientists in general, and Richard Owen in particular, confronted the new theories with snarls of dismay. Bad enough, in Owen's judgment, that evolution banished God in favor of a mechanistic picture of life. Worse yet, this dangerous doctrine had become a fad. A claim that should have inspired outrage had instead become an irresistible topic for dinner party banter and high-society chitchat.

Benjamin Disraeli, one of the sharpest observers on the Victorian scene, captured the mood. Disraeli was a compelling figure. A Jew who rose to the top of his narrow-minded society, he was also a successful and talented novelist.

His wit set him apart, and so did his appearance. He was a dandy in a dowdy age, resplendent in velvet coats and satin trousers, weighed down with rings, adorned with ringlets. He was destined to serve two terms as prime minister, and he was well aware of his many gifts. "Disraeli was a self-made man," one contemporary remarked, "and he never stopped worshiping his Creator."

A character in his 1847 novel, *Tancred*, explained the new theory that everyone was talking about. "First there was nothing, then there was something; then, I forget the next. I think there were shells, then fishes; then we came, let me see, did we come next? Never mind that, we came at last."

This was a confusing creed, and what was more confusing still is that those shells and fishes and other forms of life were all related. Disraeli's characters—and the Victorian public—tried to take it in, but the harder they tried, the more befuddled they grew.

Owen set out to crush this detestable doctrine before it could grow into dogma. Evolution was the ultimate slippery slope. If one species could change into another, then perhaps humankind was not set apart from the rest of the natural world. Perhaps, as a few daring thinkers had suggested long before Darwin, apes really were our kin!

Far more was at stake than the merit of a scientific theory. If evolution was true, the geologist Adam Sedgwick warned, then "religion is a lie; human law a mass of folly and a base injustice; morality is moonshine; . . . and man and woman are only better beasts!"

For Owen, in the 1830s and '40s, two quests had merged into one. One was to win the title of "the English Cuvier"—Cuvier himself had died in 1832—as the era's reigning authority on anatomy and paleontology. That meant Owen had rivals to dispatch. The second task was to shoot down evolution. Dinosaurs would prove the key to both ventures.

In 1837, the British Association for the Advancement of Science had awarded Owen a hefty fee—the equivalent of $160,000 today, by one historian's reckoning—to write a report on "fossil reptiles of Great Britain." The choice of Owen as the best candidate to survey and evaluate this vast field confirmed his place in the front ranks of British science.

Owen took on the assignment with his customary diligence, devoting two years to reading scientific papers and examining fossils. He delivered his talk, the first of two, in 1839. This was a long and learned dissertation on seagoing creatures like the ichthyosaurs and plesiosaurs that Mary Anning had discovered. The lecture was a hit, and Owen was acclaimed as "the greatest anatomist living."

His follow-up talk was an even bigger triumph. The subject was again fossil reptiles, but this time the focus was on land-based creatures. In 1842, in the published version of his lecture, Owen announced the existence of a group of prehistoric animals that had never been linked together. These were *dinosaurs*—Owen invented the word—and he listed the first members of the club. At this point there were only three.

One was megalosaurus, the immense carnivore that William Buckland had identified by poring over bones from the Ashmolean Museum at Oxford University. Another was iguanodon, the elephant-sized herbivore that Owen's rival Gideon Mantell had named for its iguana-like tooth. The third was lesser known, a smaller, lower-to-the-ground herbivore named hylaeosaurus, perhaps ten to twenty feet long and adorned with long spikes and a coat of armor. Mantell had discovered this creature, too, about a decade after he had found his iguanodon.

Owen's announcement was daring, for the three animals that he had marked as special seemed to have little in common. Here were, in the admiring words of Stephen Jay Gould, "a fierce carnivore, an agile herbivore, and a stocky, armored herbivore."

None of these dinosaurs looked the least bit like the dramatic skeletons in today's natural history museums. (Nearly three more decades would pass before the first dinosaur skeleton was presented to the public. That unveiling took place in 1868, at Philadelphia's Academy of Natural Sciences, and it drew such vast crowds that the academy had to build newer, larger quarters.)

In Owen's day, each of his three dinosaurs consisted only of scattered bones and teeth, a bit of jawbone here, a length of backbone there, perhaps a rib or two. These were the first dinosaurs to have a title bestowed on them, but they looked less like prehistoric titans than like fossilized roadkill.

What traits did they share? They were big and old and toothy, but so were many other creatures. What was it that made *these* three, in Owen's words, "altogether peculiar among Reptiles"?

The crucial clues, Owen informed the British Association, were a handful of anatomical features that he proceeded to list. At the base of their spines, for instance, these beasts—unlike other reptiles—had five vertebrae fused together. More conspicuously, their leg bones were thick, sturdy columns and "more or less resemble those of the heavy pachydermal Mammals." Dinosaurs' legs were far different from those of reptiles like alligators and crocodiles, which were short and stuck out sideways from the body.

These seemingly mundane traits, Owen argued, were in fact telltale. Dinosaurs were land animals, and *that* was why fused vertebrae were important. For huge, bulky animals stomping their way along, Owen pointed out, some way of strengthening the backbone would have been essential. Seagoing creatures, who were buoyed up by water, wouldn't have the same problem. For dinosaurs, fused backbones weren't a frill, like silver buttons on a dandy's coat; they were a structural necessity, like iron wheels on a locomotive.

This was new. Mantell pictured his iguanodons, for instance, as supersized lizards up to two hundred feet in length. That would have made them ten times as long as crocodiles. Owen cut that figure down to twenty-eight feet and raised iguanodons up off the ground on tall, strong legs. Mantell's iguanodon was a lizard the size of a float in Macy's Thanksgiving Day parade. Owen's iguanodon bore a greater resemblance to a gigantic rhinoceros with an oversized head and fierce, tooth-lined jaws.

Once he had set out his argument and painted his word pictures, Owen spelled out what he had achieved. In the annals of discovery, this may count as history's longest and most formal *Eureka!* "The combination of such characters," Owen wrote, ". . . all manifested by creatures far surpassing in size the largest of existing reptiles will, it is presumed, be deemed sufficient ground for establishing a distinct tribe or sub-order of Saurian Reptiles, for which I would propose the name of *Dinosauria.*"

Owen chose the name *dinosaur* with care. It came from two Greek words, he told the British Association, *deinos,* for "terrible," and *sauros,* for "lizard." Crucially, Owen meant *terrible* in its original sense of "formidable and fearsome" rather than in the modern sense of "really, really bad." (The old usage resonated with Victorians, who knew countless passages like one in Deuteronomy: "The Lord thy God is among you, a mighty God and terrible.")

Dinosaurs, in Owen's view, were majestic, awe-inspiring creatures, not loathsome, overgrown lizards. And they were, he would soon go on to argue, proof that the theory of evolution could not possibly be correct.

It is a tribute to Owen's prowess as an anatomist that, working with a comparative handful of ill-sorted bones and lacking any understanding of evolution, he managed to identify a "distinct tribe" of prehistoric creatures. (He might have done even better. Modern scientists now agree that Owen missed half a dozen dinosaurs that he could have added to his original trio.)

It might seem curious to define creatures as large as dinosaurs on the basis of traits as small as the shape of a rib or vertebra, but that's how the game is played. Mammals are defined in a similar way. Mammals are animals that have hair and suckle their young, as everyone knows, but they also have three bones in their inner ear, a particular kind of ankle joint, and a lower jaw that consists of a single bone.*

Nearly two hundred years after Owen's day, scientists can draw on mountains of evidence he never knew. Dinosaur bones have accumulated in vast numbers since the first discoveries; better still, scientists and fossil hunters have unearthed many virtually complete skeletons.

But Owen's original insight still stands up. "Dinosaurs" did indeed constitute a special category of animals, and the name tags that Owen stuck to their chests in 1842 still fit today.

---

* A definition that focused exclusively on hair and milk would miss some mammals and gather in other animals that don't belong. Some mammals are hairless or nearly hairless, like whales and dolphins. (Hair does rule an animal *into* the mammal club. No non-mammals have hair.) And some non-mammals, like spiders, feed their young on milk.

CHAPTER 40

"When Troubles Come, They Come Not Single Spies but in Battalions"

Owen soared to fame after his dinosaur talk. He had mapped the prehistoric world and planted a flag, and both the scientific community and the public hailed him as the rightful ruler of that vast and ancient territory. Owen was introduced to Queen Victoria and Prince Albert (and enlisted to tutor the royal children in science). He dined with the prime minister.

To his delight, he could not keep up with the demands on his time. An expert whose expertise seemed to have no bounds, he was urged to weigh in not only on dinosaurs, but on recent sightings of a "Great Sea Serpent," and the cause of cholera, and the philosophy of Emerson, and the paintings of J. M. W. Turner. He accumulated honors and medals by the basketful.

In the meantime Owen's rival Gideon Mantell, who had once hoped to win the "English Cuvier" title for himself, had nearly disappeared from view. This was particularly cruel, since two of Owen's three dinosaurs were Mantell's discoveries.

A smart bettor would have foreseen Owen's triumph. Mantell was a genuine expert on fossils and a tireless worker, but Owen was better connected and the superior scientist. ("Richard Owen was the greatest anatomist and paleontologist of his age," in Stephen Jay Gould's judgment. "His accomplishments were legion, both in range and excellence.")

Owen outdid Mantell, as well, in the dark arts of downplaying his rivals' discoveries and calling attention to their mistakes while highlighting his own achievements. "Owen did not see that others paved the way for him," one historian of science remarks. "Rather, they committed basic faults that he was at pains to rectify."

That made any contest between the two men a mismatch. Owen had a zest for combat and a gift for skewering his enemies; Mantell shrank from confrontation and was, in the judgment of one distinguished historian, "a clearly complex and unhappy character of an extremely easily persecuted nature."

One public clash highlighted the difference in the rivals' temperaments. At a meeting of the Royal Society on a March night in 1848, they knocked heads over the proper classification of a squidlike fossil called a belemnite. (We encountered belemnites in chapter 12, where Mary Anning and Elizabeth Philpot found a way to revive the ink from two-hundred-million-year-old belemnite fossils.)

The meeting ran late, until nearly midnight. Owen had devoted a full thirty minutes to rebuking Mantell for challenging him. Mantell staggered out of the crowded room. "I came home to my desolate hearth, suffering in mind and body," he told a friend later, "and felt how vain are all earthly pursuits." Owen went home invigorated by the battle and in carefree high spirits. He dove eagerly into the latest installment of *Dombey and Son*, the new Dickens novel, and, utterly absorbed, "stayed up very late reading."

Mantell had good reason to feel downcast. The belemnite story was only one episode in a long, sad saga. Though Mantell had made more dinosaur finds than anyone else, in the public mind dinosaurs were Owen's province.

Mantell had devoted decades of his life to fossil collecting, but he was dogged all his life by misfortune that would have left Pollyanna weeping in a corner. In 1833, he had moved his family (and his fossil hoard) to Brighton, a prosperous town on England's southern coast. He bought a grand home within sight of George IV's magnificent royal residence, put his fossils on display, and opened the doors to the public.

It didn't work out. Mantell had planned to charge visitors an admission fee, but friends had persuaded him that this would be bad form. What money he earned would have to come from his medical practice. But the life of a provincial doctor was hard and hectic, especially in these pre-anesthesia days. Mantell was perpetually running from one emergency to another, from delivering babies to setting broken arms and mending fractured skulls.

In Brighton, he hoped, he might lure well-to-do patients and ease up a bit. That hope soon fizzled. Mantell was a skilled and conscientious physician, but many prospective patients stayed away from a doctor who had newly arrived in town and whose first love, they suspected, was not medicine but geology.

In 1835, Mantell agreed to a desperate plan. In return for a fee from Brighton, his home would be converted into a full-fledged "scientific institution" that would feature not only Mantell's immense fossil collection but also a library, a reading room, and lectures. Visitors could purchase annual subscriptions (for free admission thereafter) or tickets at the door. Unfortunately, town planners pointed out, there would not be room for the Mantell family to remain in their home. But Gideon could live in a bedroom in the attic.

By this time Mary Mantell had grown weary of her husband's obsession. Turned out of her house, she took her four children and moved back to Lewes, the town they had left for Brighton.

Mary had once been her husband's ally on fossil-hunting excursions, but those days had passed. "I have no companion—no one whose smile or approbation would cheer me on," Mantell lamented in a letter to a friend. ". . . There was a time when my poor wife felt deep interest in my pursuits, and was proud of my success," he went on, "but of late years that feeling had passed away, and she was annoyed rather than gratified by my devotion to science." (In the very next sentence, as if to demonstrate that devotion, Mantell remarked that he had just made "a large collection of fossil fresh-water shells.")

The scientific institution scheme never worked out. Pressed for money and in despair, Mantell soon ended up selling his entire fossil collection to the British Museum. The work of thirty years, the collection numbered thousands of items and included some of the best dinosaur fossils ever unearthed. All gone!

On a December morning in 1838, Mantell peeked out of the attic window of his Brighton home at the commotion below. There British Museum employees wrestled crate after fossil-packed crate into horse-drawn wagons, for the journey to London. The line stretched ninety wagons long.

Richard Owen delivered his two-part report on British fossils in 1839 and 1842. In putting that survey together and claiming dinosaurs as his own, he made great use of the superb fossil collection at the British Museum, so recently the property of Gideon Mantell.

Owen and Mantell had many reasons to dislike each other, and they took advantage of all of them. Both were proud, prickly, talented men who shared an obsession with fossils. They might have clashed simply out of competitiveness. But they came to fossils from different directions, and the difference in perspective made for an extra measure of disdain and hostility.

Mantell was a collector who worked outdoors climbing up cliffs and scouting out quarries in search of hidden treasure. Owen's knowledge of fossils was unmatched, but he was a museum man who rarely went into the field.

He had once ventured out with Mary Anning as his guide. The experience only strengthened his resolve to pass up such opportunities in the future. "Next day we had a geological outing with Mary Anning and had like to have been swamped with the tide," he grumbled in a letter to his sister. "We were cut off from rounding a point and had to scramble over the cliffs."

Many professions have their own version of this hands-on–versus–hands-off, doer-versus-thinker split. All doctors know the ancient joke that "internists know everything and do nothing, and surgeons know nothing and do everything."* Newspaper reporters look at editors as desk-bound wretches who wouldn't know a story if it bit them; editors see reporters as akin to toddlers who believe that every scribble they churn out deserves a frame and a gold star.

In science the prejudice in favor of theoreticians as opposed to "mere" collectors and naturalists runs deep. If you were to believe Owen, Mantell complained, "all [Mantell] had done was to collect fossils and get others to work them out."

That view would have been unfair, but it likely did reflect Owen's judgment. Owen and Mantell lived long before the twentieth-century physicist Ernest Rutherford. Even so, they would instantly have understood his cutting remark that "all science is either physics or stamp collecting."

Owen would have nodded his head in vigorous agreement; Mantell would have fumed. *Stamp collecting indeed!* Just after Owen's "dinosaur" talk and still smarting at what he took to be Owen's poaching on his territory, Mantell sent an indignant letter to a friend. "I have again to regret a want of honour, and I may say justice," he wrote, "towards those but for whose labor and zeal he could never have obtained the materials for his own reputation."

But Mantell saw correctly how the story would play out. He would be sidelined, and Owen's intellectual gifts and imperious manner would see him installed as a scientific Admiral of the Fleet.

---

* In the full version of the joke, which leaves no specialty unmaligned, "psychiatrists know nothing and do nothing; pathologists know everything and do everything, but too late."

Return of the Happy World

O wen would do battle, first of all, with evolution. In its pre-
Darwinian form, evolution was synonymous with progress. The
story of life was a saga of steady improvement.

This was an all but universal view, a carryover from the dazzling
advances in technology that marked the new century. Giddy writers
competed to describe additional wonders that were no doubt just around
the corner; they imagined robot hairdressers, submarine ocean liners,
personal flying machines.

It was reality—especially reality in the form of roaring, hissing railroad trains—that had set the public's imagination afire. Since the dawn
of history, humankind had been subject to a strict speed limit—no one
had ever traveled faster than a horse could gallop. Now, suddenly, life
had sped up to an almost unfathomable degree.

In 1829, an Englishman named Thomas Creevey received an invitation to ride a "LocoMotive" on a five-mile test ride. He wrote about the
momentous day in his diary.

Trembling with excitement, Creevey had climbed aboard the train,
stopwatch in hand. "It is really flying," he wrote, "and it is impossible to
divest yourself of the notion of sudden death." The LocoMotive raced
down the track at twenty miles an hour and even reached a speed of

twenty-three miles an hour. Creevey climbed down afterward suffering from a headache and vowing never to repeat the experience.

Technology, it seemed, was marked by relentless advance. So was history. So, nearly everyone presumed, was the history of life. That gave Owen the opening he needed.

Dinosaurs had lived at the dawn of time, which meant—if you listened to the evolutionists—that they should have been crude and primitive organisms. Instead, they were marvels of complexity and sophisticated design. Hence, evolution was wrong. QED.

That was a welcome conclusion, because it returned God to his accustomed place of honor. Owen quoted, with admiration, his geologist friend William Buckland. The most important thing to know about dinosaurs, Buckland argued, was not that they were fearsome but that they were beautifully engineered.

"Even in those distant eras" when dinosaurs had flourished, wrote Buckland, "the same care of the common Creator, which we witness in the mechanism of our own bodies . . . was extended to the structure of creatures, that at first sight seem made up only of monstrosities."

God was on the case today, and here was proof that he had *always* been on the case.

But scientists, even ones like Owen and Buckland who took for granted God's role in governing his creation, faced the daunting question of just how he did it. Through the eons a great many species had gone extinct— that made for endless controversy, as we have seen—but that was only half the mystery. The *appearance* of new species was just as bewildering as the disappearance of old ones.

Where had new organisms come from? They hadn't all been created at once—if they had, you would have found the same fossils no matter what layer of rock you examined. So what *had* happened?

Perhaps God had fashioned each species to suit its particular circumstances. That was Owen's view at first, although he would change his mind. His original idea was that God had created each new species from scratch; contrary to what the evolutionists thought, a new species did *not* come into being because an old species had somehow transformed itself. "There was no gradation or passage of one form into another," Owen declared in 1841, "but that they were distinct instances of Creative Power, living proofs of a divine will and the work of a divine hand, ever superintending and ruling the existence of our world."

God was perpetually at work, Owen went on, tirelessly engaged in "the continuous operation of the ordained becoming of living things."

Thomas Huxley quoted Owen's words with malevolent glee. *"The ordained becoming of living things"*?! "It is obvious that it is the first duty of a hypothesis to be intelligible," Huxley wrote, "and this . . . may be read backwards, or forwards, or sideways, with exactly the same amount of signification."

Owen would later change tack. His new theory kept God in charge but seemed to make room for something that veered toward evolution. (Owen danced around this dangerous charge.) In the prehistoric past, Owen suggested, God had seeded the world with a few species and laid down laws that governed how they would change through the eons. Then he had pressed "play" and sat back contented.

This revised theory had the great virtue of keeping God at center stage, but there was something distressingly vague about those divine laws. What exactly were the rules of change that God had set down?

Owen admitted that he could not say. "To what natural laws or secondary causes the orderly succession and progression of such organic phenomena may have been committed we as yet are ignorant," he declared in the closing passages of an 1849 lecture.

But in the next sentence, he delivered a rousing finale. But if it was true that we did not yet understand the laws that God had set in place to govern the natural world, that was not the essential point. We would understand them someday. It was the *existence* of the rules that was crucial.

With rules in operation, chance played no role in the story of life. Each species took its turn stepping onstage in accordance with the divine plan. "Generations do not vary accidentally, in any and every direction," Owen proclaimed, "but in preordained, definite, and correlated courses."

God had decreed, at the dawn of time, how the story of life would play out. From humble beginnings, life would rise to great heights, and that long climb would culminate in nature's greatest achievement, humankind!

Owen framed that message in the grandly murky language he favored. His words require deciphering today, but to his Victorian listeners they were as inspiring as a hymn. "Nature has advanced with slow and stately steps," Owen proclaimed, "guided by the archetypal light, amidst the wreck of worlds, from the first embodiment of the Vertebrate idea . . . until it became arrayed in the glorious garb of the Human form."

This was an old message draped in a new and splendid cloak. Here was the much-maligned "happy world" back again, and this time not merely with a philosopher's say-so but presented as an impossible-to-question assertion of scientific doctrine.

The reference to "archetypal light" was obscure but key. Archetypes were "Divine ideas," Owen wrote, blueprints for the design of animal bodies that God reused time and again. Look at the skeletons of a mouse and a horse and an elephant and a dinosaur, Owen explained, and you would see variations on a single theme—here, in one creature, were the bones that formed the spinal column; here, in another, were bones that looked the same, or nearly so, arranged in nearly the same way.

Owen looked closer. In meticulous detail, he compared the bones in a whale's flipper, a mole's paw, and a bat's wing. Each "hand" had the same five fingers; each finger had the same three joints. That was surely not coincidence.

Each forearm matched "bone for bone," Owen noted, and so did each upper arm. The bones were not identical—a bat's wing was "sustained, like an umbrella, by slender rays," and everything that was "elongated and attenuated in the bat was shortened and thickened in the mole"—but the essential design was unchanged.

Owen followed up with a curious observation. Naturalists sometimes called animals "living machines," he noted, but animals differed in a crucial way from machines. A boat, which was built to cut through the water, had a completely different design from a tunnel-boring machine, which was meant to dig through the dirt. Neither had any resemblance to a hot-air balloon, which was made to drift across the sky. That made sense, considering their different jobs. But whales and moles and birds, though built for the different jobs of swimming, digging, and flying, shared the same design.

Why?

Darwin would soon provide an answer—common design was proof of common ancestry. Owen proposed a different answer. Where Darwin saw a family resemblance, Owen saw an artist who never tired of replaying his greatest hits.

Why wouldn't the Creator have favorite themes and a recognizable style? That is, after all, how artists work. Vermeer repeatedly painted solitary women absorbed in thought, in tall rooms with light pouring in from a window on a left-hand wall. Dickens over and over again created orphans who, we learned at last, were not penniless waifs but misplaced children of the upper crust.

Owen's picture of life unfolding in line with a divine plan was a characteristically Victorian hybrid of science and religion. Here was a scientific theory that was also, in the words of Loren Eiseley, a "strange spiritual drama," a kind of opera played out on a global scale and across countless eons.

In Eiseley's brisk paraphrase of Owen's obscure language, whole species appeared and then vanished, while "the great patterns of life, the divine blueprints, one might say, persisted from one age to another." God had settled on his blueprints at the beginning of time, Owen contended, and he had envisioned how they would be modified through the ages. This was true of all forms of life whatsoever and especially true of the highest form of all, humankind.

"The knowledge of such a being as Man must have existed before Man appeared," Owen wrote. "For the Divine mind which planned the Archetype also foreknew all its modifications."

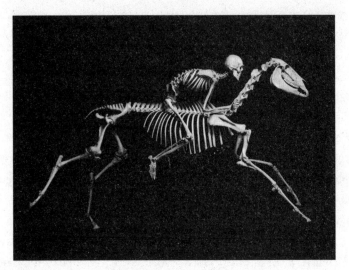

A human rider atop a horse.
How to explain the similarities in their skeletons?

Owen had *almost* solved the greatest mystery of the age, and he believed he had done a great thing. He had devised a sweeping theory that explained the history of life as a grand procession headed to a pre-ordained goal. He had, it would soon turn out, gotten things profoundly wrong, as Huxley and other mockers would delight in pointing out.

But for the moment, all was well. Owen was at center stage, in effect, basking in applause, conspicuous in the spotlight, turning this way and that to acknowledge his admirers. He took a step forward and bent into an especially deep bow. He didn't yet see that he was about to topple into the orchestra pit.

Dinner in a Dinosaur

F ew shocks stay shocking, and in Britain in the 1840s and '50s the
science-minded public soon grew accustomed to talk of immense
fossils and endless eons. Even in Tennyson's circles, familiarity blunted
fear. This is often the way, and not just in science. In the art world a few
decades later, for instance, critics responded to the first Impressionist
exhibitions with sputtering fury. Only "lunatics" could have produced
such "obscenities." Eventually those despised works grew beloved, and
the hideous daubs were reproduced by the millions in posters and cal-
endars. Other cultural landmarks were greeted in similar fashion. The
Eiffel Tower was denounced, at first, as "a truly tragic street lamp." A full
three decades after the premiere of Beethoven's Fifth Symphony, one
prominent critic still condemned its "odious meowing."

In the case of science, as we have seen, the taming process began
with presenting discoveries in a way that drained away their strangeness. Biblical "days" were ages. "In the beginning" was a reference to
the moment when humans first stepped on the scene. *Nothing to see here,
people. Keep moving!*

Still, there were dissenters. As they saw it, scientists had done their
best to declare God irrelevant. They had tried to disguise their misdeeds;
they had given God a big desk and a corner office. But no one was fooled.

"It is mean and miserable," the Reverend Edward Henry Carr declared angrily, "to suppose that God our father has actually left us to ourselves— left us to be tossed up and down by the blind play of natural laws."

But those indignant voices were drowned out by a larger, more hopeful chorus. Earlier generations had hailed God's skill as a designer and proclaimed his special fondness for humankind; their successors had *demonstrated* the truth of those claims.

By the early 1850s, scientists were in an exultant mood. Geologists and paleontologists could point to half a century of accomplishment. Starting from nowhere—the word *geology* did not even exist until 1795, and *paleontology* not until 1833—scientists had racked up triumph after triumph. They had discovered reams of fossils, they had resurrected creatures beyond counting, they had flung open the gates of time, they had fought down their own religious doubts and dispatched their fundamentalist rivals.

The time had come to celebrate.

Few dinner parties have ever matched the razzle-dazzle of Benjamin Waterhouse Hawkins's bash on New Year's Eve 1853. The guest list was starry. The menu was elaborate. But it was the setting that was unprecedented.

Hawkins had invited his distinguished guests to gather in a one-of-a-kind spot—inside a huge, cut-open model of a dinosaur! Twenty-two of Britain's most prominent scientists, and a handful of highly placed editors and other eminent men (and no women), took their places around two banquet tables.

The dinosaur—a life-sized model of an iguanodon—was so big and stood so high that the servers carrying their silver platters had to climb a staircase to reach the diners. The beast's back had been cut away, to make room for the tables. Richard Owen, who shared the evening's top billing with Hawkins, sat at the head of the main table (and at the iguanodon's head, as well).

Benjamin Hawkins was a renowned artist and sculptor and more than a bit of a showman. He'd been commissioned to build full-sized replicas of some three dozen prehistoric creatures, including three dinosaurs, for an immense exhibition that was due to open in June in South London, six months after the New Year's banquet.

Hawkins had lavished special care on his dinosaurs. (The iguanodon alone stretched thirty-five feet and weighed thirty tons.) The unveiling of the sculptures, at a grand opening presided over by Queen Victoria and Prince Albert, would mark the first time the public had ever laid eyes on dinosaurs as they might have been.

The party was held in Hawkins's studio, a vast space crowded with immense dinosaurs under construction, and huge, froglike amphibians, and oversized quasi-turtle reptiles. A pink and white tent encircled the diners (and helped ward off the winter chill). Chandeliers and candelabras cast a flickering glow. Banners high above the tables carried the names of the stars of the dinosaur saga, including Owen, Cuvier, Buckland, and Mantell.

Hawkins had planned every aspect of the evening's celebration with meticulous care. He had personally drawn each invitation, with slight variations from guest to guest. One invitation featured an immense iguanodon, with gaping mouth and fearsome claws (and a cutaway back depicting dinner guests seated at a table). Alongside the iguanodon, a diner clung to the long neck of one of Mary Anning's plesiosaurs and waved his wineglass in the air.

A pterodactyl with an upraised wing stood front and center. The wing carried a handwritten message: "Crystal Palace. Mr. B. Waterhouse Hawkins requests the honor of _____'s company at Dinner, in the Iguanodon."

The whole evening was a tribute to Victorian excess, from the diners' formal dress to the enormous sculptures that jammed the studio to the eight-course feast itself, which featured such dishes as mock turtle soup and *côtelettes de mouton aux tomates* and *turbot à l'hollandaise*, and moved on to charlotte russe and an array of other desserts, and then grapes and pears and almonds and walnuts, all washed down by rivers of sherry and madeira and port and Moselle and claret.

Benjamin Waterhouse Hawkins's studio. The iguanodon model,
complete with misplaced horn on its nose, is in the center.

The festivities began at four in the afternoon and continued into the new year, with endless rounds of tipsy toasts and songs and rambling, self-congratulatory speeches.

At about the eight-hour mark, near midnight, a geologist guest named Edward Forbes launched into a lengthy poem in honor of the iguanodon that had housed them for the evening.

> A thousand ages underground
> His skeleton had lain;
> But now his body's big and round,
> And he's himself again!

Forbes continued on for several more verses and then invited everyone to join in the chorus.

> The jolly old beast
> Is not deceased,
> There's life in him again.

The diners roared along "in a manner so fierce and enthusiastic," one journalist reported, "as almost to lead to the belief that a herd of iguanodons were bellowing."

If Hawkins and Owen had been able to peek six months into the future, they might have sung one more happy, rowdy chorus. The Crystal Palace Dinosaur Park was a colossal hit. On opening day, forty thousand dazzled Londoners turned up to gawk at the prehistoric monsters.

Day after day, week after week, the crowds poured in. Special trains had to be rushed into service. Two million visitors a year gaped at the dinosaurs, for decade after decade.

The dinosaur park was a kind of sequel to one of the greatest triumphs of the age, the Crystal Palace Exhibition, which had been held in London's Hyde Park in 1851. The "Great Exhibition of the Works of

Industry of All Nations," as it was formally known, was the first-ever world's fair. Its most striking feature was the Crystal Palace itself, an enormous and innovative glass-and-iron greenhouse that housed thousands upon thousands of exhibits in light-flooded splendor.

When the fair ended, the Crystal Palace was taken apart, transported a few miles away to South London, and rebuilt. Hawkins's dinosaurs had been commissioned as star attractions for the new site.

The sculptures occupied a landscape of their own, complete with islands in man-made lakes, where the dinosaurs struck fierce, lifelike poses. (Hawkins took advantage of the lakes by half-submerging some of his models when he lacked crucial information about a particular bit of anatomy.) Isolated bones in museums, even immense bones, had been nothing like this. Illustrations in scientific articles didn't compare. Here was a new world that everyone—not only scientists and other specialists—could visit.

"The sensational statues acted on viewers in the same way frescoes affected early Renaissance worshippers," writes the historian Zoë Lescaze. "Just as Giotto rendered the life of Christ viscerally real and immediate to illiterate congregations half-listening to Latin sermons they could not understand, Hawkins vividly conveyed prehistory to broad audiences."

A modern photo of Hawkins's Crystal Palace dinosaurs

The sculptures represented state-of-the-art scientific thinking. Owen and Hawkins had worked closely together. The dinosaurs looked almost mammal-like, in line with Owen's views, and not at all like Mantell's overgrown lizards.

By today's standards, the models look off-target in a host of ways. (Their odd appearance shows, the historian of science Nicolaas Rupke points out, that the talk about deducing an entire creature from a single bone was wildly overblown.) Modern scientists picture iguanodons, in particular, as more agile than Hawkins's rhinoceros-like beasts and more poised to rise up on their hind legs. And Hawkins's iguanodon sported a large horn on its nose, also like a rhinoceros. The horn was misplaced, it now seems, in pin-the-tail-on-the-donkey fashion. Scientists now think the horn was really a thumb, a sharp spike that made a good weapon. (Owen thought so, too, but apparently Hawkins disagreed.)

Dinosaurs had nothing to do with the "works of industry" that the Crystal Palace had been built to showcase, but that didn't matter. In Victorian eyes, the massive sculptures were as worthy of celebration as the steam engine or the locomotive or any other emblem of modernity and power. Hawkins's dinosaurs were "an expression of successful conquest," in the judgment of the historian Martin Rudwick, a fitting part of "the imperial celebration that characterized the Crystal Palace exhibits as a whole."

The public's logic here was not airtight, but no one looks for logic in a toast. Dinosaurs were dead and we weren't, and that was excuse enough to lift a glass to the story of triumphant humanity.

In any case, the English knew all about "imperial celebration." They'd had the inspired idea, for instance, of commemorating Wellington's victories over Napoleon by making a statue from melted-down Napoleonic cannons. Now they exulted over an earlier generation of fallen foes.

London's *Quarterly Review* put the boast in taunting form, in an article celebrating the New Year's Eve banquet. "Saurians, Pterodactyls all! . . . Dreamed ye ever . . . of a race to come dwelling above your tombs and dining on your ghosts."

*   *   *

Certainly Hawkins and Owen and their fellow diners were in a jovial mood on New Year's Eve 1853. Long after midnight, when the last toast had been delivered and the last bottle emptied and the last chorus sung, the celebrants left for home. It was a raw night, snowy and so cold that, for the first time in more than ten years, the Thames was choked with ice. But when the dinner guests stumbled their cheery way into the dark, no one paid much heed to the weather.

They had reason to feel complacent. After some confusing years when dinosaurs had crashed the party and time had expanded dizzyingly, scientists had managed to restore order to their world.

The story had come full circle. Half a century before, the world had been inviting and orderly. Then things had fallen apart. But dinosaurs had been vanquished, the corridors of time had been tidied up, and all was quiet and cozy once again.

Most important of all, humankind still occupied its old, exalted position. The story of life on Earth was grand and uplifting—perhaps even more so in this new telling than in the past—and humans stood front and center. Everything led up to us.

No conclusion seemed more solid. But if you listened hard enough— if, for a moment, you managed to ignore the diners' happy bellowing— then perhaps you could just make out a faint ticking sound. It was almost as if someone had planted a bomb.

"It Is Like Confessing a Murder"

The bomb maker was Charles Darwin, perhaps history's most reluctant revolutionary. Darwin was a Victorian gentleman with a comfortable income, a large home and handsome grounds, and a dread of controversy. Shy to the verge of reclusiveness, Darwin ducked nearly every invitation that came his way. He was so protective of his privacy that he had the road in front of his house dug deeper and a six-foot-high wall built so that passersby could not peek into his windows.

Darwin was, in every area except science, a man of enlightened but conventional views. No one could have been less like a café philosopher in Paris, sipping espresso and holding forth on the meaninglessness of life. And yet Darwin did more than any world-weary existentialist to undermine the notion that life's hardships were part of a divine plan and that our suffering and striving serve a deep and vital purpose.

Darwin's subject was the natural world as a whole, not human life in particular. But for the New Year's Eve partygoers at the dinosaur banquet, and for countless old-school Victorian intellectuals, Darwin's rise meant *their* fall.

As soon as Darwin stepped into the open, the cozy assumptions that they had long endorsed—that the world was orderly and presided over by a benevolent deity who had a special fondness for humankind—were blasted into rubble.

After Darwin, no one would ever again portray natural history as a saga that built steadily and inevitably upward to an act 3 coronation of humankind. The history of life, Darwin showed, was less a noble progression than a drunken stagger.

Darwin detonated his bombshell, *On the Origin of Species*, in 1859. The basic idea was simple. There are not enough seats at the table. Every living organism has no choice but to join in a high-stakes game of musical chairs. With so many competitors, even the smallest advantage—ever so slightly sharper elbows, a smidgen quicker reflexes—might prove crucial. And since the game goes on forever, round after round, any advantage that is inherited might ratchet up, and in time the descendants of the first competitors might come to look vastly different from their ancestors.

Real life is more complicated, because the environment can change from year to year. (In effect, the number of chairs can change, and so can the music.) What had been a good strategy in wet years might do poorly in times of drought. A predator might invade, like the snakes that stowed away in the landing gear of airplanes and found easy prey in Guam's lizards and birds. A watering hole might be paved over or a forest chopped down.

The theory sounds straightforward, but what made it controversial was what it implied, not what it said outright. Species *do* change, Darwin explained, but not by setting their sights on some far-off goal. They change in response to the here and now. The winnowing and shaping process that Darwin called "natural selection" might give us glamorous peacocks or, just as likely, hideous parasitic worms. No outcome was preordained.

Humans play no particular role in the story. We came along, just as brontosauruses and bacteria and swans and horses came along, and there was nothing more special about our story than about theirs. As it happened, humans appeared and then spread across the globe. But if some of our long-ago ancestors had drowned in a flood or burned in a fire, we might never have turned up in the first place.

Crucially, the system ran by itself, guided by the few simple rules of musical chairs ("Everyone must play," "The music never stops," and so on). No one was in charge. In particular, God wasn't in charge, or even in the picture.

Critics howled, as Darwin had known they would. In 1844, years before he went public with the *Origin*, he had written a letter to a friend that hinted at his theory of evolution. "It is," he wrote, "like confessing a murder."

Richard Owen, perhaps the best-known scientist in England until Darwin claimed the title, was one of those screaming "Murder!" His views on evolution changed during his long lifetime and were never quite clear—historians still debate just where he stood—but through the decades Owen clung to the view that humankind stood atop the pyramid of life.

Owen was, moreover, *the* undisputed expert on fossil bones, so his views carried enormous weight. (When Darwin returned from his five-year voyage on the *Beagle*, with tons of fossil bones he had collected, it was Owen he enlisted to examine his finds and identify what was what.) Owen knew far more about fossils than Darwin or anyone else. "What differentiates revolutionary thinkers from non-revolutionary ones is almost never a greater knowledge of the facts," writes the historian of science Frank Sulloway. ". . . Darwin, who knew less, somehow understood more."

Learned as he was, Owen got evolution wrong, and that was the sort of mistake you don't recover from. Owen had a far more imposing manner than Darwin. Even so, it was modest, hesitant, soft-spoken Darwin who came up with, in the words of the philosopher Daniel Dennett, "the single best idea anyone has ever had."

What Darwin had seen was that there *could* be design without a designer. He explained in detail how that worked, but Owen and a great many others could not grasp it. (Part of the problem is that "If there is design, there must be a designer" sounds as if it *has* to be true, simply by virtue of what the words mean.)

One early critic thundered, in indignant capital letters, that Darwin's "fundamental principle" was that "IN ORDER TO MAKE A PERFECT AND BEAUTIFUL MACHINE, IT IS NOT REQUISITE TO KNOW HOW TO MAKE IT."

The hostile writer had it exactly right. That was precisely Darwin's claim. No wonder the world had such trouble believing him. No wonder so many still do.

It was surprise enough that there could be design without a designer. Darwin had a second revelation to spring on the world. Once again Owen and the other sages of Victorian times responded in two stages. First they failed to see it, and then they rushed to denounce it.

Everyone had long acknowledged that nature can be harsh. No one had any trouble picturing the grim reaper wielding his scythe to lop off the heads of the weak and sickly. What was nearly impossible to see, before Darwin, was that those death-dealing swipes could also yield new forms of vibrant life.

The point was so important that Darwin highlighted it in the final, majestic paragraph of the *Origin*. "From the war of nature," he wrote, "from famine and death, the most exalted object which we are capable of conceiving, namely the production of the higher animals, directly follows."

A century later the poet Robinson Jeffers made the point succinctly:

> What but the wolf's tooth whittled so fine
> The fleet limbs of the antelope?
> What but fear winged the birds, and hunger
> Jewelled with such eyes the great goshawk's head?

In the pantheon of scientific heroes, Darwin has a permanent place alongside such titans as Galileo and Newton. His onetime rival Richard Owen has been banished to history's attic.

When Darwin relegated Owen to permanent second-class status, you might have guessed that dinosaurs, which Owen had "invented," would fall along with him. Just the opposite happened.

Dinosaurs grew more glamorous than they had ever been, especially in the decades around 1900 when magnificent new skeletons turned up in the American West. Robber barons and magnates rushed to sponsor expeditions and build natural history museums. In our day, billionaires strive to outdo one another by shooting rockets into space. A century ago they competed to bring back a bigger, fiercer T. Rex skeleton than anyone had ever seen.

Visitors have rushed to see those finds ever since, in museums and movie theaters across the globe. The creatures on exhibit have changed over the decades, but the fascination today remains what it was in the beginning.

In 1854, an anonymous visitor made his way to Hawkins's Crystal Palace Park to see the dinosaurs. He was drawn, he wrote, by the chance to peek into "a dim world, where monsters dwell."

Epilogue

Nearly all the players in the dinosaur saga met hard ends. Posterity has been harsh to Owen—in science, even more than in other fields, the winners write the history books—but he was a complicated man, and two things were true at once. Owen was brilliant, and he was impossible.

Impossibly vain, for one thing. His response to a request for biographical information, late in his life, was characteristic. "The peaceful career of this indefatigable cultivator of Natural Knowledge," he began (referring to himself in the third person), "has been a continued series of labours for the promotion of scientific truth and its practical application to the well-being of mankind."

But Owen's "peaceful career" was in fact marked by bitter feuds, and at a time when he needed allies, he found himself alone. Even mild-mannered Charles Darwin, a man notoriously hard to provoke, made an exception for Owen. "I used to be ashamed of hating him so much," Darwin complained to a friend after Owen had sought to undermine a rival, "but now I will carefully cherish my hatred & contempt to the last day of my life."

Owen's most enduring legacy is the Natural History Museum in London, a magnificent cathedral to nature. Owen lobbied hard for the creation of a new museum, separate from the British Museum, and he served as its first director. The museum is one of the world's architectural treasures (and, to this day, it is free and open to all, as Owen insisted). It boasts vast open spaces, stained-glass windows, towering walls topped with beautifully rendered sculptures depicting a huge variety of animals,[*] and a collection of tens of millions of objects, with dinosaurs prominent among them.

A statue of Owen once occupied one of the most conspicuous settings in the museum, across from the main entrance. Today Owen has been deposed. In his place sits a larger-than-life statue of Charles Darwin.

On the top floor of the Natural History Museum, incidentally, discreet signs indicate "The Anning Rooms." Here donors and other dignitaries gather to discuss business or celebrate special occasions.

A portrait of Mary Anning hangs on the wall. So do framed, handwritten letters she wrote describing her fossil finds. On another wall a large message carries the inspiring but almost certainly irrelevant words "She sells seashells by the seashore . . ." The rooms are quiet and elegant, the sort of setting where riffraff like Mary Anning herself would never have been seen.

Few of the other players in our story received even the posthumous recognition that came Mary Anning's way. William Buckland, the genial, eccentric soul who was the first to identify a dinosaur, died raving mad in an asylum. Richard Owen lived until 1892, but he was one of those for whom the name died before the man.

Gideon Mantell's story was especially dark. Fate seemed to have had nothing better to do than to torment Mantell, who suffered one calamity after another. The fossil hunter/physician went to his deathbed

---

[*] The dozens of animals that perch high on the museum's outside walls range from lions and eagles to saber-toothed tigers and pterosaurs. Few visitors recognize one marker of Owen's anti-evolutionary views—in keeping with his belief that species did not change, Owen kept extinct animals and living ones separate. All the extinct species stare out from the museum's east wing; the living beasts preside over the western wing.

convinced that Owen had stolen credit for *his* finds and that the world had endorsed the theft.

We looked on, earlier, as Mantell watched vans line up to carry away the fossils he had gathered over a lifetime. By then, his wife had left him, in fury and frustration. Still to come was the long, painful illness of his favorite daughter, Hannah, who died of tuberculosis at age eighteen.

A year after Hannah's death, in 1841, Mantell was hurrying to see a patient when the coachman let the reins of his horse get tangled up. Mantell leapt out to unsnarl the mess while the coach was still moving. He crashed to the ground and hit his head, and then the wheels grazed his head.

Mantell had long been afflicted with what is now called scoliosis, a curvature of the spine. Two weeks after his coach accident, he wrote that he was "ill with symptoms of paralysis, arising from spinal disease."

Despondent and in pain, he wrote in his journal, "Could I choose my destiny, I would gladly leave this weary pilgrimage." Six months later, he recorded a new journal entry and a new woe: "Tumour of considerable size has gradually made its appearance on the left side of the spine."

The year of these medical crises, 1841, happened to be the year when Owen told the world about dinosaurs. He took all the credit for himself, Mantell complained bitterly, and slighted Mantell's role in the saga.

Precisely as Mantell had feared, the world hoisted Owen onto its shoulders to hail his dinosaur discoveries. Mantell was left to mutter angrily on the sidelines.

Mantell suffered on for another decade, much of it trapped in a sickbed. He never fully regained his strength, but his pain ebbed and flowed, and in the good stretches he took up his work. Then his symptoms would return and he would do his best to dose himself with bed rest and brandy and opium.

Mantell died in 1852. His twisted spine ended up in a formaldehyde jar, as a medical curiosity illustrating "the severest degree of deformity of the spine." The jar sat on a shelf in the Hunterian Museum, in London, the very museum run by Mantell's archenemy, Richard Owen. (The story is sometimes told as if Owen himself had sought out Mantell's spine, as the capstone to a malevolent career. But it was Mantell who had suggested that his spine be examined after his death and preserved at the Hunterian if it proved unusual.)

Only one story—the story of the dinosaurs themselves—had a happy ending. Happy in comparison, at any rate. Dinosaurs will be famous forever, first of all, and, what is more important, they were granted an enviable finale. Dinosaurs reigned unchallenged for an unimaginable one hundred million years. Then, in a cataclysm that reverberated around the globe, with no warning, no foreboding, no lingering, they vanished.

Acknowledgments

I had, once upon a time, a collection of plastic dinosaurs that any of my fellow six-year-olds would have recognized immediately. The yellow and green beasts battled one another and starred in crayon-drawn epics I scribbled on large sheets of poster board. It never dawned on me to wonder in those early years—not at the time and not for decades afterward—when it was that people had first learned about these astonishing animals.

Every nonfiction book is built around a question. Mine was, What was it like when scientists and ordinary men and women first learned that the world once contained lizards as big as elephants?

The search for an answer would stretch to several years, most of it devoted to happy wandering through archives and a far briefer stint to clambering over fossil-rich cliffs. Fortunately, the nineteenth century was a letter-writing, diary-keeping era. With a bit of digging—I did far better hunting for letters and memoirs than for fossils—I was able to hear voices that last spoke one hundred and fifty years earlier. I'm deeply grateful to a host of librarians and archivists who helped guide me along. The most fruitful hunting grounds were the treasure-rich Natural History Museum in London, and the American Museum of Natural History in New York, and such tiny but beautifully curated gems as the Lyme

Regis Museum, built on the site of Mary Anning's home in Dorset, England, and the Charmouth Heritage Coast Centre, a few miles away, on what is now known as England's Jurassic Coast.

Walter Jahn, a biologist at the State University of New York, and Lee Alan Dugatkin, a biologist and historian at the University of Louisville, helped me resolve a long string of mysteries. I turned often, too, to Phil Davidson, an English fossil hunter who retains the exuberance of the dinosaur-loving schoolboy he once was. I owe a special debt to Dr. Richard Carlson, a geologist whose knowledge is matched only by his patience. Carlson is a staff scientist emeritus at the Carnegie Institution for Science, in Washington, DC, and he interrupted his own busy days time and again to help a stranger who seemed never to run out of questions.

Colin Harrison shaped and sharpened a story that roamed across continents and millennia. Emily Polson provided thoughtful counsel and unfailing support. Aja Pollock is the ideal copyeditor, equally at home fixing a typo and recasting a shaky argument. Flip Brophy is my agent and, more important, my friend. She is a wise and loyal ally, savvy and unshakable in her enthusiasm.

My two sons are both writers. It is an unusual day that I do not tap their expertise. Books are long hauls, and they are the ideal companions on those demanding journeys.

Lynn deserves more thanks than I can put in words.

Notes

Sources for quotations and for assertions that might prove elusive can be found below. To keep these notes in bounds, I have not documented facts that can be quickly checked in standard sources. Publication information is provided only for books and articles not listed in the bibliography.

Epigraph

ix *"How does the once unthinkable"*: Jack Gross, "Historicizing the Self-Evident: An Interview with Lorraine Daston," *Los Angeles Review of Books*, January 5, 2020. Online at https://tinyurl.com/3ct5y46z.

Introduction: A Shriek in the Night

1 *"It is a happy world"*: Paley, *Natural Theology*, 456. Online at https://tinyurl.com/2n2e4avn.

1 *"the grandest and most precious"*: Ibid., 139.

2 *"Birds of prey"*: Ibid., 481.

2 *"The world was made"*: Ibid., 456.

2 *"was second only to that"*: Sulloway, "Intelligent Design."

3 *"The first world war"*: John D. Ray, *The Rosetta Stone and the Rebirth of Ancient Egypt* (Cambridge, MA: Harvard University Press, 2007), 25.

3 *"The lesser nations"*: White, *Atrocities*, 261.

3 *Two hundred crimes*: Morrison, *Regency Years*, 22–23.

5 *On an ordinary day*: Hitchcock, *Ichnology of New England*, 3, and Gilette and Lockley, eds., *Dinosaur Tracks and Traces*, 11.

7 *"crush beneath your carriage-wheels"*: This was William Buckland, speaking to the British Association for the Advancement of Science, in 1836. Quoted in Richardson, *Sketches*, 16.

9 *the word* scientist: Clark, *Bugs and the Victorians*, 11.

Chapter One: "Dragons in their Slime"

11 *"Dragons of the prime"*: Tennyson, "In Memoriam AHH," canto 57.

13 *"a real bone"*: Plot, *Natural History of Oxfordshire*, 132.

13 *"It will be hard to find"*: Ibid., 131.

14 *"those great Tusks"*: Ibid., 135.

14 *"Notwithstanding their extravagant Magnitude"*: Ibid., 136.

14 *"Goliath for certain"*: Ibid.

16 *signs of musculature*: Flannery, "Dinosaur Crazy."

Chapter Two: The Girl Who Lived

17 *The Girl Who Lived*: For admirers of J. K. Rowling everywhere, the story of Mary Anning and the lightning strike will immediately call Harry Potter to mind. I decided to stick with the chapter title "The Girl Who Lived" even after I found that Mary Anning's best biographer, Tom Sharpe, had gotten there before me, in chapter 2 of *The Fossil Woman: A Life of Mary Anning*. Sharpe's biography is remarkable for its thoroughness and careful scholarship.

18 *Only Mary and an older brother*: Torrens, "Mary Anning," 258.

18 *"left by the mother"*: Sharpe, *Fossil Woman*, 24.

19 *Jane Austen was one*: Emling, *Fossil Hunter*, Kindle location 177.

20 *"a closing-over element"*: Adam Nicolson, *Why Homer Matters: A History* (New York: Henry Holt, 2014), 68.

20 *Even in the Middle Ages*: Ritchie, *Lure*, 12.

20 *"the ocean is a wilderness"*: Ibid. The passage is from Thoreau's *Cape Cod*. Herman Melville disagreed. He began *Moby-Dick* with a long, lyrical passage about the delights of staring at the ocean and daydreaming. "Meditation and water are wedded forever," Melville wrote, but his was distinctly a minority view.

21 *"a Kind of common Defense"*: Russell, *Dissertation on the Use of Seawater*, vi.

21 *Well-off travelers*: Blei, "Inventing the Beach."

22 *(a cure for deafness)*: Sharpe, *Fossil Woman*, 109.

22 *Mary began helping out*: Marie-Claire Eylott, "Mary Anning: The Unsung Hero of Fossil Discovery," Natural History Museum (UK). Online at https://tinyurl .com/2te2x23x.

23 *"Many swim in"*: Scott Cleary, "The Ethos Aquatic: Benjamin Franklin and the Art of Swimming," *Early American Literature* 46, no. 1 (2011): 53.

24 *"The country people here"*: Philip Wayne, ed., *Letters of William Wordsworth* (New York: Oxford University Press, 1954), 21.

25 *"having fallen over a cliff"*: Torrens, "Mary Anning," 259.

Chapter Three: "The Most Amazing Creature"

27 *"fully determined to go"*: Sharpe, *Fossil Woman*, 33.

27 *"I found [the Annings]"*: Torrens, "Mary Anning," 261.

28 *"An eye," one nineteenth-century scientist said*: Buckland, *Geology and Mineralogy*, 138.

28 *"the tyrant of the deep"*: Glendening, "Ichthyosaurus," 24.

29 *a local landowner paid*: Lescaze, *Paleoart*, 18.

29 *"altogether the most monstrous"*: Torrens, "Mary Anning," 264.

30 *"a serpent threaded through a turtle"*: Gideon Mantell, *Medals of Creation*, 595, paraphrasing a remark of the geologist Adam Sedgwick. (The description is often misattributed to William Buckland.) Mantell's two-volume work is online at https://tinyurl.com/npfx28zv.

30 *"To the head of a Lizard"*: Torrens, "Mary Anning," 264.

30 *So unlikely was the creature*: Ibid.

31 *("A man of devouring ambition")*: Edey and Johanson, *Blueprints*, 29.

31 *"the most monstrous assemblage"*: Torrens, "Mary Anning," 264, and Louis Figuier, *Earth Before the Deluge*, quoted in Rudwick, *Scenes from Deep Time*, 188.

32 *"We adjourned to the Society's rooms"*: McGowan, *Dragon Seekers*, 72. The colleague was Henry De la Beche, who had worked with Conybeare at spotting the first plesiosaur bones.

32 *ten men had spent a day*: Sharpe, *Fossil Woman*, 67.

32 *"They were as alien"*: McGowan, *Dragon Seekers*, 72.

Chapter Four: An Epic Written in Chalk

34 *To the poet John Burgon*: Hugh Kenner, *The Pound Era* (Berkeley, CA: University of California Press, 1971), 124.

35 *Take the White Cliffs of Dover*: Author interview with Richard Carlson, a geologist at the Carnegie Institution for Science.

35 *"countryside made of skeletons"*: Halliday, *Otherlands*, xvii.

36   *A mega-flood with a surge*: Quirin Schiermeier, "The Megaflood That Made Britain an Island," *Nature*, July 18, 2007.

36   *"A great chapter of the history"*: Thomas Huxley, "Chalk."

37   *Two of the most eminent scientists*: Tattersall, *Rickety Cossack*, 33.

Chapter Five: "The Dreadful Clink of Hammers"

39   *"Man stood at the center of all things"*: Eiseley, *Darwin's Century*, 153.

40   *(Can you tell what good the wind does)*: Rev. T. Wilson, *Lessons on Natural Philosophy, for Children* (London: Darton and Clark, 1846), 28.

40   *"Does it not seem strange"*: Miller, *Life and Letters*, 393.

40   *"fluttered in weak rags"*: Wedderburn and Cook, eds., *Works of John Ruskin*, 115.

41   *"the whole physical creation"*: Henry Ward Beecher, *Evolution and Religion* (New York: Fords, Howard, and Hulbert, 1885), 115. Beecher's book was published well after *On the Origin of Species*, but Beecher continued to hold the non-Darwinian view that evolution was synonymous with progress.

42   *"Remember you are mortal"*: The classics scholar Mary Beard examines the truth and the origins of the story in *The Roman Triumph* (Cambridge, MA: Belknap Press, 2009), 85–92.

Chapter Six: "It's a Beautiful Day and the Beaches Are Open"

43   *"myriads of 'creeping things'"*: Mantell, "Reptiles," 182. Everyone in the nineteenth century would have recognized that "creeping things" was a biblical allusion. On the fifth day of Creation, God said, "Let the earth bring forth the living creature after his kind, cattle, and creeping things, and beast of the earth after his kind. And it was so."

43   *"the fleshless bones"*: Hawkins, *Memoirs*, 59.

44   *"The Fiend . . . With head, hands"*: In his *Scenes from Deep Time* (pp. 68–69), Martin Rudwick discusses this passage from Buckland and quotes *Paradise Lost*.

44   *"shrieked against a straight reading"*: Brooke, *Science and Religion*, 226, quoted in Freeman, *Victorians and the Prehistoric*, 5.

44   *euphemisms like* fallen angel: See, for instance, Sally Mitchell, *The Fallen Angel: Chastity, Class, and Women's Reading, 1835–1880* (Bowling Green, OH: Bowling Green University Popular Press, 1981).

45   *"seemed not so much reality"*: Rudwick, *Scenes from Deep Time*, 62.

45   *"on the slimy shores"*: Percy Bysshe Shelley, "Prometheus Unbound," act 4.

46   *dead, black-button eyes*: Robert Shaw makes the point in unforgettable, hammy fashion as Quint the shark hunter in *Jaws*: "You know the thing about a shark, he's got lifeless eyes, black eyes, like a doll's eyes. When he comes at you, he doesn't seem to be living. Until he bites you, and those black eyes roll over white."

Chapter Seven: Trembling in the Dark

47 *(three kinds of cone cells)*: Yong, *An Immense World*, 85.

48 *the melanosomes in dinosaurs*: Benton, *Dinosaurs Rediscovered*, 9–12.

48 *Ancient dragonflies had the wingspan*: Both these examples are from the science writer and paleontologist Riley Black, who has written many of the clearest and most vivid accounts of prehistoric life. Black, "Why Is Life Today So Small?"

48 *"Could T. Rex have bitten"*: Benton, *Dinosaurs Rediscovered*, 282.

49 *75 percent of all the species*: Preston, "The Day the Dinosaurs Died."

49 *Some of the rocks*: Katherine Kornei, "A New Timeline of the Day the Dinosaurs Began to Die Out," *New York Times*, September 10, 2019.

49 *Air temperatures reached five hundred degrees*: Black, *Last Days*, 73.

50 *"Many heads grew up without necks"*: Dugatkin and Bergstrom, *Evolution*, 29.

Chapter Eight: The Divine Calligrapher

53 *(it was Whewell who coined)*: "William Whewell," Zalta and Nodelman, eds., *Stanford Encyclopedia of Philosophy*. Stanford University compiled this online-only encyclopedia. See https://plato.stanford.edu/entries/whewell/.

54 *"lamps lighting nothing"*: Gillispie, *Genesis and Geology*, 207.

55 *poems hailing God's handiwork*: Medawar and Medawar, *Aristotle to Zoos*, 71.

55 *The precise dimensions of King Solomon's Temple*: Richard Westfall, *Never at Rest: A Biography of Isaac Newton* (Cambridge, UK: Cambridge University Press, 1980), 346–8.

56 *never uttered the word* God: Ted Davis, "The Faith of a Great Scientist: Robert Boyle's Religious Life, Attitudes, and Vocation," BioLogos, August 8, 2013. This is the site founded by Francis Collins, Nobel laureate and former director of the National Institutes of Health.

56 *"The two great books"*: Boyle, *Excellence of Theology*, 40.

56 *"Flourishes on the Capital Letters"*: Boyle, *Disquisition*, 258.

57 *"the Eye of a Fly"*: Ibid., 43. Robert Boyle and Walt Whitman were worlds apart, but passages in *Leaves of Grass* read like pages from Boyle. "I believe a leaf of grass is no less than the journey-work of the stars," Whitman wrote in *Song of Myself*, section 31, and the first stanza of that section concludes, "And a mouse is miracle enough to stagger sextillions of infidels."

57 *"A book requires an author"*: Wootton, *Invention of Science*, 444.

58 *"What savage fierceness"*: John Wesley, "The General Deliverance," 1872. Online at https://tinyurl.com/yakjk82t.

Chapter Nine: The Apple of God's Eye

59   *"to kill a pagan"*: St. Bernard of Clairvaux, quoted in James Reston Jr., *Warriors of God: Richard the Lionheart and Saladin in the Third Crusade* (New York: Anchor, 2002), 12.

59   *"Beliefs held almost without question"*: This is the first sentence of Casey, *After Lives*.

60   *"The saints not only do not celebrate"*: Origen, *Homilies on Leviticus 1–16*, trans. Gary Wayne Barkley (Washington, DC: Catholic University of America Press, 1990), 156.

60   *"The God that holds you"*: Jonathan Edwards, "Sinners in the Hands of an Angry God. A Sermon Preached at Enfield, July 8th, 1741," University of Nebraska Electronic Texts in American Studies, 15. Online at https://tinyurl.com/2p9jep77.

60   *"The hinges in the wings"*: Paley, *Natural Theology*, 541.

60   *"Why add pleasure"*: Ibid., 484.

60   *"but in moderate drops"*: Ibid., 371–3.

61   *"If we suppose a wise"*: Whewell, *Astronomy*, 31. This was the third in the series called *The Bridgewater Treatises on the Power, Wisdom, and Goodness of God as Manifested in the Creation*. Whewell was one of eight leading thinkers who delivered a Bridgewater lecture. Online at https://tinyurl.com/46k72prx. Whewell's comment spurred mild-mannered Charles Darwin to indignant sputtering. Whewell had been praised as "profound because he says length of days adapted to duration of sleep in man!! whole universe so adapted!!! & not man to Planets. instance of arrogance!!" See *The Transmutation Notebooks of Charles Darwin, Notebook D*, online at https://tinyurl.com/3erubhes.

61   *"the human race is just a chemical scum"*: From an interview with Ken Campbell in "Beyond Our Ken," a 1995 episode of docuseries *Reality on the Rocks* (Channel 4 Television, UK).

62   *"Christians are like a council of frogs"*: Greenblatt, *Swerve*, 98.

62   *"All the parts of the universe have me in view"*: Ibid., 246.

Chapter Ten: Whales in the Treetops

66   *Could a clam, say*: John Moore, *Science*, 103.

66   *"There is no such thing as shell"*: Cutler, *Seashell*, 134.

66   *"Why should not Nature"*: Ray, *Discourses*. Online at https://tinyurl.com/2ykfxt67.

67   *Hooke built a spectacular collection*: Jardine, *Curious Life*, 37–38.

67   *"Why does she not imitate"*: Hooke, *Earthquakes*, 145.

67   *"Bones, horns, teeth, and claws"*: Ibid.

67   *Hooke was a devout Christian*: Jardine, *Curious Life*, 89.

67   *"And 'tis not unlikely also"*: Hooke, *Earthquakes*, 75.

67 *Fossils were "written in more"*: Rudwick, *Earth's Deep History*, 68, from a 1668 lecture to the Royal Society.

68 *"a piece about the bigness"*: Hooke, *Micrographia*, 107.

68 "petrifying *water (that is, such a water)*": Ibid., 109.

69 *"where birds had twittered"*: Thomas Huxley, "Chalk."

69 *"foolish and simpleminded"*: Isaacson, *Leonardo*, 439.

70 *"the bottom of the sea was raised"*: Ibid., 440.

70 *"imaginary influences"*: Hooke, *Earthquakes*, 147.

70 *("placed there by Merlin")*: Rappaport, "Hooke on Earthquakes," 141.

70 *no one paid much attention*: Ibid., 129.

70 *"It is much more natural to suppose"*: Voltaire, "The Changes That Have Happened in Our Globe," in *The Works of Voltaire*, vol. 19, *Philosophical Letters* (E. R. Dumont, 1733). Online at https://tinyurl.com/5srkpruc.

Chapter Eleven: Without a Trace

73 *Near reefs, bodies can vanish*: Mark Ridley, "Dreadful Beasts," *London Review of Books*, June 28, 1990.

74 *Out of every eighty million Tyrannosaurus rexes*: Kenneth Chang, "How Many Tyrannosaurus Rexes Ever Lived on Earth? Here's a New Clue," *New York Times*, April 16, 2021.

76 *professional collectors tell stories*: Author interview with Phil Davidson, October 19, 2022.

76 *"It's a race against time"*: Author interview with Phil Davidson, November 16, 2022.

77 *"enormous bone weathering out"*: McGowan, *Dragon Seekers*, 88.

77 *"The old horror movie trope"*: McKay, *Mammoth*, 2.

77 *"There are countless stories"*: Author interview with Phil Davidson, December 21, 2022.

77 *Under assault by acid*: Nigel R. Larkin, "Pyrite Decay: Cause and Effect, Prevention and Cure," *NatSCA News* 21 (2011), 35–37. (This is a publication of the UK-based Natural Sciences Collections Association.)

77 *Ninety-nine percent of all the animal species*: Williams, *Dinosaur Artist*, xvi.

77 *"the students far outnumber"*: Coyne, *Why Evolution Is True*, 197.

Chapter Twelve: "None of the Advantages"

79 *"arguably the first person"*: Goodhue, "Origins." Online at https://tinyurl.com/yu33arw8.

79 *"This persevering female"*: Torrens, "Mary Anning," 261.

79 *"The Cliff fell upon me"*: Sharpe, *Fossil Woman*, 124.

79 ("*For the last year*"): Ibid., 103.

80 "*brought fainting from the beach*": Ibid., 118.

80 "*I was so intent*": Ibid., 97.

80 "*In one place we had to make haste*": Ibid., 63.

80 "*Mary untangled the seaweed*": Ibid., 125.

81 "*she had one of her miraculous escapes*": Ibid., 124.

81 "*The Word of God is becoming precious*": Ibid., 125.

82 "*Her faith let her do*": Goodhue, "Origins."

82 *inherited a sugar plantation*: Sharpe, *Fossil Woman*, 42.

82 *(his collection weighed twenty tons)*: Rudwick, *Scenes from Deep Time*, 64.

82 *the first two years of their marriage*: Geikie, *Murchison*, 87.

82 *By about age eleven*: Goodhue, "Origins."

82 "*She was the right person*": Ibid.

82 *(even when those papers discussed fossils)*: Sharpe, *Fossil Woman*, 71.

82 *copy and original are often*: McGowan, *Dragon Seekers*, 16.

83 *(She took a break)*: McGowan (*Dragon Seekers*, 15) and Sharpe (*Fossil Woman*, 124) tell slightly different versions of the story. (They differ a bit on timing and on just which Conybeare paper Anning was copying.)

84 *Mary Anning's name went unrecorded*: Sharpe, *Fossil Woman*, 136, and Torrens, "Mary Anning," 280–81.

84 ("*She were good to the poor*"): Sharpe, *Fossil Woman*, 117.

84 "*therefore it is my intention*": Ibid., 80.

84 "*Mary Anning takes the liberty*": Ibid., 81.

84 "*She says the world has used her ill*": Sharpe, *Fossil Woman*, 115, and Davis, "Mary Anning," 108.

84 *(So important were those finds)*: Rupke, *Great Chain*, 135.

85 *Anning would continue on*: Torrens, "Mary Anning," 266, cites five major finds.

85 "*In short, a monster*": Sharpe, *Fossil Woman*, 88–89.

85 "*the industry and skill*": Ibid., 85, 87, 91.

86 "*the celebrated Miss Anning*": These were the words of the geologist Charles Lyell, quoted in Sharpe, *Fossil Woman*, 69.

86 "*the Princess of Paleontology*": This was a remark by the German explorer Ludwig Leichhardt, who visited Anning in 1837. Quoted in Torrens, "Mary Anning," 269.

86 "*this poor, ignorant girl*": Torrens, "Mary Anning," 265.

Chapter Thirteen: "Sister of the Above"

87 "*sallied out [to Lyme Regis]*": Torrens, "Mary Anning," 268.

88 "*And what is a woman?*": Sharpe, *Fossil Woman*, 123.

88 *"noticed by all the cleverest men"*: Sharpe, *Fossil Woman*, 115.

88 *"The fact is that I am going to sell my collection"*: Ibid., 54.

88 *(a doubly good deed)*: Dean, *Mantell*, 58.

88 *"how I wish you could see it"*: The letter is online at https://tinyurl.com/2jm9wjwz.

89 *"Exuberant, chaotic, and revolutionary"*: Lescaze, *Paleoart*, 17.

89 *"a gentleman whom she had entrusted"*: Sharpe, *Fossil Woman*, 128.

90 *"It wasn't a fortune"*: Ibid., 129.

91 *The story turns up repeatedly*: See the careful debunking essay by the folklorist Stephen Winick, "She Sells Seashells and Mary Anning: Metafolklore with a Twist," writing at the Library of Congress website, July 26, 2017. Online at https://tinyurl.com/2dv7ttcs.

91 *"I have never been out of the smoke of Lyme"*: Sharpe, *Fossil Woman*, 95.

Chapter Fourteen: Ferns and Fox Hunters

95 *"the weirdest walk-on"*: Kathryn Schulz, "The Ten Best Weather Events in Fiction," *New Yorker*, November 20, 2015.

95 *"ballerinas of doom"*: Michael Greshko, "A Tyrannosaur Was Found Fossilized, and So Was Its Last Meal," *New York Times*, December 8, 2023.

97 *"The noble science of fox-hunting"*: Geikie, *Murchison*, 93.

97 *"he was an independent gentleman"*: Ibid., 129.

98 *"We are here in a divine country"*: David McCullough, *Brave Companions: Portraits in History* (New York: Simon and Schuster, 1992), 6.

98 *"a mighty river flowing"*: Mantell, *Sussex*, 51.

99 *"a Swiss army knife"*: Wallace, *Beasts of Eden*, 4.

100 *(Historians now question)*: Dean, *Mantell*, 72.

100 *Mantell wrote disconsolately*: Cadbury, *Terrible Lizard*, 89.

101 *"mementos of wretchedness"*: Ibid., 100.

Chapter Fifteen: Into the Temple of Immortality

103 *"Might we not have here"*: Dean, *Mantell*, 81.

103 *more than ten thousand skeletons*: Moore, *Knife Man*, 237.

103 *Hunter was reputedly the model*: Wallace, *Beasts of Eden*, 19.

103 *"a pile-up of genitals"*: Asma, *Stuffed Animals*, 87.

105 *"one of the most gigantic reptiles"*: Mantell, *Sussex*, 78.

105 *"two years earlier they had sneered"*: Dean, *Mantell*, 84.

105 *"I shall ride on the back"*: McGowan, *Dragon Seekers*, 87.

106 *"the virtually unanimous disapproval"*: Dean, *Mantell*, 84.

106 *"To call Mantell 'the discoverer'"*: O'Connor, *Earth on Show*, Kindle location 296.

106 *"I have made a grand discovery"*: Dean, *Mantell*, 111.

107 *He named it hylaeosaurus*: Ibid., 113.

Chapter Sixteen: Framed for Bliss

109 *"Like Frankenstein," he wrote*: Mantell, *South-east of England*, 315.

109 *"perpetual warfare and incessant carnage"*: Buckland, *Geology and Mineralogy*, 131.

109 *"No country on the face"*: Mantell, *Wonders*, 501.

110 *"The whole class of herbivora"*: Buckland, *Geology and Mineralogy*, 132.

111 *"The feeble and disabled are speedily relieved"*: Ibid, 131.

111 *He dubbed his affliction*: Chapman, *Caves, Coprolites and Catastrophes*, 8.

111 *the sketchy medical accounts of the day*: "William Paley," in *Dictionary of National Biography*, vol. 43 (London: Smith, Elder & Co., 1895). Online at https://tinyurl.com/3jjaty3x.

111 *After three years of torment*: Chapman, *Caves, Coprolites and Catastrophes*, 32.

112 *a "sublime truth"*: Mantell, *Wonders*, 504.

112 *"Th' earth's face is but thy table"*: John Donne, "Elegy on Mistress Bulstrode."

112 *"the whole earth . . . is cursed and polluted"*: Gale, "Darwin."

112 *"one great slaughter-house"*: Ibid.

113 *("a well-balanced, harmonious system")*: Ibid.

113 *John Maynard Keynes once remarked*: Keynes, *The General Theory of Employment, Interest and Money* (London: Macmillan, 1936), viii.

113 *"Ere man was called into existence"*: Mantell, *Wonders*, vol. 2, 504.

113 *"Fragments vast of lost creations"*: Richardson, *Sketches*, 6–7.

Chapter Seventeen: "A Delicate Toast of Mice"

115 *a fetish of eccentricity*: James Gregory, "Eccentric Biography and the Victorians," *Biography* 30, no. 3 (Summer 2007).

115 *"not behavior that at once commands"*: Allen, *Naturalist*, 56.

115 *his first peek at Buckland*: Gordon, *Buckland*, 10.

116 *"heard the animal munching"*: Tim Fulford, "Romancing the Stone: Coleridge and Geology," *Coleridge Bulletin*, New Series 37 (September 2011), 47.

116 *fox, rabbits, ferrets, hawks*: Gordon, *Buckland*, 102.

117 *"point out and explain"*: Ibid., 30.

117 *"undergroundology"*: Ibid., 40.

117 *Was all this wealth "mere accident"*: Ibid., 82–83.

117 *a long essay in the* Christian Review *(fn.)*: "History and Destiny of Coal," *Christian Review* 21 (April 1856), 282.

118 *"When we see the body of an Ichthyosaurus"*: Buckland, *Geology and Mineralogy*, 157.

118 *"He paced like a Franciscan preacher"*: Gordon, *Buckland*, 31.

119 *Hedgehog was "good and tender"*: Ibid., 104.

119 *"I have always regretted"*: John Ruskin, *Praeterita: The Autobiography of John Ruskin* (1885; repr., Oxford, UK: Oxford University Press, 1975), 192.

119 *The worst thing he had ever tasted*: Augustus Hare, *The Story of My Life*, vol. 5 (London: George Allen, 1900), 358.

119 *"while the guinea-pig under the table"*: William Tuckwell, *Reminiscences of Oxford* (London: Cassell, 1901), 39.

119 *"I can tell you what it is"*: Ibid., 40.

119 *"Dr. Buckland, whilst looking at it, exclaimed"*: Hare, *Story of My Life*, 358.

120 *straight out of a rom-com*: Gordon, *Buckland*, 91.

120 *("Rather more than two years")*: Geikie, *Murchison*, 86.

120 *Mary kept a journal*: Gordon, *Buckland*, 92.

120 *kept up their scientific collaboration*: Tuckwell, *Reminiscences*, 37.

121 *William fetched their tortoise*: Ibid.

Chapter Eighteen: Kirkdale Cave

123 *"The bottom of the cave"*: Gordon, *Buckland*, 60.

124 *like pigeon legs through a pie crust*: McGowan, *Dragon Seekers*, 53.

124 *Buckland's first thought*: Rupke, *Great Chain*, 32.

125 *"Scarcely a single bone"*: Gordon, *Buckland*, 61.

125 *remnants of two hundred to three hundred hyenas*: Ibid., 62.

126 *"You have just made a discovery"*: Greene, "Genesis and Geology," 139.

126 *Oxford was almost as much*: Ibid., 144.

126 *"Science teaching was not intended"*: Secord, *Victorian Sensation*, 223.

127 *"Billy has performed admirably"*: Gordon, *Buckland*, 58–59.

127 *(Buckland had planned to finish up)*: Ibid., 57.

128 *More good news came from India*: The scientist was Charles Lyell, writing to Gideon Mantell. See *Life, Letters and Journals of Sir Charles Lyell, Bart.*, ed. K. S. Lyell, vol. 1, (London: John Murray, 1881), 164.

129 *never before gone to a geologist*: O'Connor, *Earth on Show*, Kindle location 1627.

130 *it was those glaciers*: Stephen Jay Gould discusses Buckland's shift in thinking in an essay called "The Freezing of Noah," in *The Flamingo's Smile*.

130 *Tim Flannery has even suggested*: Flannery, "Dinosaur Crazy."

130 *"amongst outrageous madmen"*: Gruber and Thackeray, *Richard Owen*, 79.

Chapter Nineteen: In Nature's Cathedral

131 *The whole country seemed possessed*: Gates, "Natural History," 540.

132 *"just as competitive as cricket"*: Desmond and Moore, *Darwin*, 57.

132 *"a presence that disturbs me"*: The lines are from Wordsworth's "Lines Composed a Few Miles Above Tintern Abbey."

132 *"a walk in the country"*: Aldous Huxley, "Wordsworth in the Tropics." Online at https://tinyurl.com/urbd8u95.

132 *("It droppeth as the gentle rain from heaven")*: David Landes quotes Shakespeare in this context in *The Wealth and Poverty of Nations: Why Some Are So Rich and Some So Poor* (New York: Norton, 1999), 17.

132 *"Nature abounded in 'wonderful adumbrations'"*: Conniff, *Species Seekers*, 207.

132 *"Almost everyone—scientist, novelist"*: Shortland, "Darkness Visible."

132 *"Wondrous shapes, and tales terrific"*: This was George Richardson, quoted in *Excavating Victorians* by Virginia Zimmerman, 31.

134 *(It's not true that Victorians)*: Matthew Sweet, *Inventing the Victorians: What We Think We Know About Them and Why We're Wrong* (New York: St. Martin's Press, 2001), xii.

134 *"not only by having living young"*: Barber, *Heyday*, 133.

134 *larding a poem (about a princess's adventures)*: Alfred, Lord Tennyson, "The Princess," in *Tennyson: A Selected Edition*, ed. Christopher Ricks (Berkeley, CA: University of California Press, 1989), 264.

134 *Tuesday mornings were for chemistry*: Meadows, "Astronomy and Geology."

134 *"The hills are shadows"*: Tennyson, "In Memoriam AHH," canto 124.

135 *"scientists, poets, and artists felt"*: Hughes-Hallet, *Immortal Dinner*, 149.

135 *"No one was barred"*: Gates, "Natural History," 541.

135 *Aristocrats competed for the chance*: Kenyon, "Science and Celebrity."

135 *"Tickets for the lectures"*: Hughes-Hallet, *Immortal Dinner*, 152.

135 *"Ladies of rank and fashion"*: Cadbury, *Terrible Lizard*, 216.

Chapter Twenty: "Quite in Love with Seaweeds"

137 *"in consequence of an invitation"*: Christabel Rose Coleridge, *Charlotte Mary Yonge: Her Life and Letters* (1903; repr., London: Forgotten Books, 2018), 45.

138 *"This was doubtless a fatiguing operation"*: Allen, *Naturalist*, 118.

139 *"quite in love with seaweeds"*: J. W. Cross, ed., *Life of George Eliot as Related in Her Letters and Journals* (New York: Crowell, 1884), 204.

139 *a dress in a seaweed pattern*: Ann Christie, "A Taste for Seaweed: William Kilburn's Late Eighteenth-Century Designs for Printed Cottons," *Journal of Design History* 24, no. 4 (2011), 306.

139 *Victoria made a seaweed album*: Thad Logan, *The Victorian Parlour: A Cultural Study* (Cambridge, UK: Cambridge University Press, 2001), 124, and Carol Armstrong and Catherine de Zegher, eds., *Ocean Flowers: Impressions from Nature* (Princeton, NJ: Princeton University Press, 2004), 111.

140 *One entrepreneur's story offers some idea*: Barnett, *Sound of the Sea*, 194–95.

Chapter Twenty-one: William Paley Stubs His Toe

141 *A mole could race*: John Leonard Knapp, *The Journal of a Naturalist* (Philadelphia: Carey & Lea, 1831), 106.

142 *"Suppose I had found a watch"*: Paley, *Natural Theology*, 1.

142 *"Every indication of contrivance"*: Ibid., 17.

142 *Cicero had looked at sundials*: Dennett, *Darwin's Dangerous Idea*, 29 fn.

142 *seemed even more compelling*: Karen Armstrong makes this point in *The Case for God* (New York: Knopf, 2009), 228.

143 *"It is unnecessary to dwell"*: Paley, *Natural Theology*, 346.

143 *the world had been created by "some infant deity"*: David Hume, *Dialogues Concerning Natural Religion* (1779), part 11. Online at https://tinyurl.com/yzn5j4r7.

144 *"To walk where you are walking"*: Stephen E. Hunt, "'Free, Bold, Joyous': The Love of Seaweed in Margaret Gatty and Other Mid-Victorian Writers," *Environment and History* 11, no. 1 (February 2005), 20.

144 *studied nature out of "blind curiosity"*: Thompson, *Note-book*, 12.

144 *a one-size-fits-all philosophy*: O'Connor, *Earth on Show*, Kindle location 1152, and Brooke, *Science and Religion*, 287–88.

144 *"The Victorians saw nothing glorious"*: Barber, *Heyday*, 26.

Chapter Twenty-two: Here Be Dragons (and Giants and Cyclopses)

147 *"A given genius may come either too early or too late"*: William James, *The Will to Believe* (1896), 230. Online at https://tinyurl.com/2p9ububv.

148 *"Iguanodon, a three-ton herbivorous dinosaur"*: Lescaze, *Paleoart*, 27.

148 *a six-ton sculpture of a fierce dragon*: Mayor, *First Fossil Hunters*, 34.

150 *"A principal courtyard of Whitehall Palace"*: Peter Mason, *Infelicities: Representations of the Exotic* (Baltimore, MD: Johns Hopkins University Press, 1998), 68.

150 *"As late as 1789"*: Cohen, *Fate of the Mammoth*, 27.

150 *In India, oversized bones*: Ibid., 132, citing Hugh Falconer, *Paleontological Memoirs and Notes of the Late Hugh Falconer, with a Biographical Sketch of the Author*, vol. 1 (London: Hardwicke, 1868), 43.

150 *a Roman historian told*: Mayor, *First Fossil Hunters*, 145.

151   *"the gloomiest of men"*: Pliny the Elder, *Natural History* (Cambridge, MA: Harvard University Press, 1938), 17.

151   *"The bones which are from time to time discovered"*: Cohen, *Fate of the Mammoth*, 23, quoting Saint Augustine, *The City of God*, book 15. The Augustine passage is online at https://tinyurl.com/482puw4n.

151   *"So that we could see how huge"*: McKay, *Mammoth*, 131–32.

152   *the tracks had been made by Divine Lucky Rhinoceros*: Mayor, *Flying Snakes*, 160.

152   *In medieval Germany, legends told*: Ibid., 155.

Chapter Twenty-three: Looking into Medusa's Eyes

153   *an Austrian paleontologist named Othenio Abel*: Abel was an ardent Nazi as well as a noted scientist. In 1944, when the Nazis still held power, he was recommended for the Goethe Prize on the grounds that he was "always in the front ranks in the fight against the threat of Jewishness and foreign infiltration at the University of Vienna." See Klaus Taschner, "Othenio Abel, Kämpfer gegen die 'Verjudung' der Universität," *Der Standard*, October 9, 2012. (The title means: "Othenio Abel, Fighter Against the 'Judaification' of the University.") Online at https://tinyurl.com/36wynktb.

153   *pointed out something odd*: Mayor, *First Fossil Hunters*, 35.

155   *"Unlike the other monsters"*: Mayor, *Flying Snakes*, 16.

155   *"an area inhabited by griffins"*: Mayor, *First Fossil Hunters*, 30–31.

155   *these bones were creamy white*: The paleontologist Peter Dodson makes this point in his foreword to Mayor's *The First Fossil Hunters*, p. xvii.

156   *"strewn over the surface"*: Mayor, *First Fossil Hunters*, 40.

156   *"Instead of a poorly drawn sea monster"*: Ibid., 159.

Chapter Twenty-four: Leibniz's Unicorn

159   *In northern Germany, Gottfried Leibniz wrote*: Ariew, "Leibniz on the Unicorn," 278.

160   *"In the same way the ancients"*: Ibid., 267.

160   *He decided to begin*: Ibid., 277.

160   *"ennobled our era by his discoveries"*: Ibid., 278.

160   *They snorted and strained*: Didier Maleuvre, *The Horizon: A History of Our Infinite Longing* (Berkeley, CA: University of California Press, 2011), 118.

161   *"extracted by pieces"*: Aries, "Leibniz on the Unicorn," 279.

162   *Modern-day scientists believe*: Cohen, *Fate of the Mammoth*, 45, and McKay, *Mammoth*, 38.

164   *roaring crowds watched them killed*: Jerry Toner, *The Day Commodus Killed a Rhino: Understanding the Roman Games* (Baltimore, MD: Johns Hopkins University Press, 2015), 1–2.

164 *In 1515 the sultan of Cambay*: Hoare, *Albert and the Whale*, 31.

165 *"the famous woodcut stood"*: Sue Prideaux, "How Dürer Shaped the Modern World," *New Statesman*, April 14, 2021. This is a review of Philip Hoare's *Albert and the Whale*.

165 *"For people in the early nineteenth century"*: O'Connor, *Earth on Show*, Kindle location 1760.

166 *"The pterodactyl is not represented"*: Ibid., Kindle location 1761.

Chapter Twenty-five: "The Grinders of an Elephant"

168 *they found heaps of bones*: Browne, *Charles Darwin*, 351.

169 *"a rodent similar to an elephant"*: McKay, *Mammoth*, 47.

170 *"a great prodigious Tooth"*: Stanford, "Giant Bones," 47–48.

170 *"the tooth of a Giant"*: Ibid.

170 *"the Monster was judged above 60 or 70 feet"*: Ibid., 53.

170 *Most people shared Cornbury's view*: Ibid., 48.

170 *Here was an "Illustrious Confirmation"*: Morris, "Geomythology," 710.

171 *"Men, who were able to turn the world"*: Ibid.

171 *an enslaved African man named Onesimus* (fn.): Kathryn S. Koo, "Strangers in the House of God: Cotton Mather, Onesimus, and an Experiment in Christian Slaveholding," *Proceedings of the American Antiquarian Society* 117, April 2007.

171 *"At a place in Carolina called Stono"*: Semonin, *American Monster*, 86.

172 *"In my opinion they could be no other"*: Ibid.

172 *"It appears that Negro slaves"*: Simpson, "Beginnings," 134.

Chapter Twenty-six: "The Terror of the Forest"

173 *Here were fitting symbols*: Semonin, "Peale's Mastodon."

174 *"Forests were laid waste at a meal"*: Ibid.

174 *"With the agility and ferocity of the tiger"*: Ibid.

174 *"He loved money and became rich"*: Gould, "Man Who Invented Natural History."

175 *"There are a bare five"*: Antonello Gerbi, *The Dispute of the New World: The History of a Polemic, 1750–1900* (Pittsburgh: University of Pittsburgh Press, 1973), 30.

175 *"Sloths are the lowest form"*: Gould, "Man Who Invented Natural History."

175 *twenty thousand spectators lined the streets*: Roger, *Buffon*, 433.

175 *"shriveled and diminished"*: Dugatkin, "Buffon, Jefferson."

176 *"It is also much smaller"*: Gerbi, *Dispute of the New World*, 3.

176 *"This elephant of the New World"*: Ibid., 4.

176 *dogs could not summon the strength*: Dugatkin, *Mr. Jefferson and the Giant Moose*, Kindle location 367.

176 *"In the savage, the organs"*: Dugatkin, "Buffon, Jefferson."

176 *"five or six times" bigger than an elephant*: Jefferson, *Notes on the State of Virginia*, 43.

176 *In 1781 he'd written a note*: Barrow, *Nature's Ghosts*, 17.

177 *he kept a collection of incognitum bones*: Thomson, "Jefferson's Old Bones."

177 *mammoth bones in the East Room*: Conniff, *Species Seekers*, 93, and Simpson, "Beginnings," 155.

177 *"the Bones of a Tremendious animal"*: "Megalonyx Jeffersonii Fossils," Thomas Jefferson Encyclopedia, Monticello.org. Online at https://tinyurl.com/25z75c5b.

Chapter Twenty-seven: "The Very Extraordinary SKELETON"

179 *"The claws of an African lion"*: Boyd, "The Megalonyx, the Megatherium," 424.

179 *Megalonyx was "more than three times as large"*: Jefferson, *Memoir*, 251.

180 *The fixes were subtle*: Ibid., 251.

180 *he sometimes regretted that he had chosen*: Coonen and Porter, "Thomas Jefferson and American Biology," 746.

180 *Jefferson was obsessed with data*: Ibid.

182 *Buffon's "very degrading" theory*: Richard Conniff, "When Thomas Jefferson Visited Yale," *Yale Alumni Magazine*, May/June 2009. Online at https://tinyurl.com/3pzzz379.

182 *(it meant "earth burrower")*: Prothero and Schoch, *Horns, Tusks, and Flippers*, 164.

182 *The mammoth on display*: O'Connor, *Earth on Show*, Kindle location 663.

183 *(Peale had mounted a mouse's skeleton)*: Personal communication with Lee Alan Dugatkin, author of *Behind the Crimson Curtain: The Rise and Fall of Peale's Museum*, September 29, 2022.

183 *"In the inverted position of the tusks"*: Semonin, "Peale's Mastodon."

184 *"In the present interior of our continent"*: Jefferson, "A Memoir," 69.

184 *"the animals of the country generally"*: Thomas Jefferson, "Instructions for Meriwether Lewis, 20 June 1803." Online at https://tinyurl.com/muawxhnm.

Chapter Twenty-eight: Noah's Ark

185 *no European had ever seen a kangaroo*: Andrea Hope, "Artwork in Focus—Sydney Parkinson, Kangaroo, 1770," AustralianArtHistory.com, May 24, 2022. Online at https://tinyurl.com/3p4trz9x.

186 *"a thousand kinds of birds and fowls"*: Poskett, *Horizons*, 17.

186 *"European culture experienced something"*: Greenblatt, *Marvelous Possessions*, 14.

186 *"the English had only cod"*: Grigson, *Menagerie*, 5.

186 *sentenced to seven years' transportation*: Bethan Bell, "The Rise, Fall, and Rise of the Status Pineapple," BBC News, August 2, 2020.

187 *Not all royals were thrilled (fn.)*: Grigson, *Menagerie*, 9.

187 *It was "pestered with visits"*: "Stubbs' Zebra," Huntington Library, Art Museum, and Botanical Gardens exhibition information. Online at https://tinyurl .com/3t99uvvx.

187 *Satirists churned out bawdy songs*: Christopher Plumb, "'The Queen's Ass': The Cultural Life of Queen Charlotte's Zebra in Georgian England," in *The Afterlives of Animals: A Museum Menagerie*, ed. Samuel J. Alberti (Charlottesville, VA: University of Virginia Press, 2011), 24, and H. Howard, "The Queen's Ass," Broadside Ballads Online from the Bodleian Libraries.

187 *"the most beautiful creatures ever seen"*: Grigson, *Menagerie*, 73.

187 *If a creature was strange enough*: Stanley Finger and Marco Piccolino, *The Shocking History of Electric Fishes: From Ancient Epochs to the Birth of Modern Neurophysiology* (New York: Oxford University Press, 2011), 282.

188 *After the eel came a baboon*: Grigson, *Menagerie*, 92, 104.

188 *in Galileo's words, it was "often very abstruse"*: Galileo Galilei, "Letter to the Grand Duchess Christina of Tuscany," 1615, in Internet Modern History Sourcebook, 5. Online at https://tinyurl.com/3ud49p8w.

189 *Scholars calculated the precise dimensions*: Cohn, *Noah's Flood*, 40.

190 *Carl Linnaeus had tallied some fourteen thousand species*: Browne, "Noah's Flood," 133.

190 *(even his own name eluded him)*: Broberg, *Man Who Organized Nature*, 388.

190 *"All these two hundred and eighty [species of] Beasts"*: Walter Raleigh, *The History of the World, in Five Books* (London: Robert White, T. Basset, et al., 1677), 66. Online at https://tinyurl.com/yf5hc4ek.

191 *how had they found their way back home*: Enenkel and Smith, eds., *Zoology in Early Modern Culture*, 99.

191 *"Why so many thousand strange birds"*: Cohn, *Noah's Flood*, 41.

191 *"How America abounded with Beasts of prey"*: Ibid.

191 *The key was the lost continent of Atlantis*: Browne, "Noah's Flood," 119.

192 *The two lions would devour the two sheep*: Ibid., 137.

192 *"a period in which there were lizards"*: Miller, *Testimony of the Rocks*, 192.

Chapter Twenty-nine: "A Cold Wind out of a Dark Cellar"

193 *A serious scientist and a well-rounded scholar*: Rev. Gideon Smales, "The Reverend George Young, DD," in *Whitby Authors and Their Publications* (Whitby, UK: Horne and Son, 1867).

194 *"a very curious fossil"*: Young, "Account of a Singular Fossil Skeleton."

194 *the particular species of ichthyosaur*: Ellis, *Sea-Dragons*, 68.

194 *"As the science of Natural History enlarges"*: Young, "Account of a Singular Fossil Skeleton."

194 *"Ichthyosaurs have been extinct for 93 million years"*: Ellis, *Sea-Dragons*, 68.

194 *"I trust I have shown you"*: Rupke, *Great Chain*, 220.

195 *"have their station in the subterranean waters"*: Ibid.

195 *"All this evidence, I think cannot fail"*: Bigelow, "Bigelow on the Sea Serpent," 150.

195 *"His head was rather larger"*: Ibid., 151.

195 *two eminent geologists agreed*: The two were Robert Bakewell and Benjamin Silliman. See Robert Bakewell, *An Introduction to Geology* (London: Longman, 1838), 362.

196 *a "dismemberment of the universe"*: Cutler, *Seashell*, 138.

196 *"The hint of extinction"*: Eiseley, *Firmament of Time*, 43.

196 *"Death was the first mystery"*: Fustel de Coulanges, *The Ancient City* (1864; repr., Boston: Lee and Shepard, 1901), 29.

Chapter Thirty: Sherlock Holmes Ponders a Bone

199 *his first public lecture*: Kolbert, "Lost World."

199 *"Considerable differences have long been noted"*: Cuvier, "Memoir on the Species of Elephants," 19.

200 *(he was the youngest member)*: Le Guyader, *Geoffroy Saint-Hilaire*, 12.

200 *the best natural history collection*: Rudwick, *Earth's Deep History*, 105.

200 *French scholars followed the army*: Rudwick, *Bursting the Limits of Time*, 360.

200 *the museum's collection would grow fourfold*: Le Guyader, *Geoffroy Saint-Hilaire*, 13.

201 *"Cuvier could correctly describe"*: Arthur Conan Doyle, "The Five Orange Pips" (1891).

201 *Every animal was an intricate*: Cuvier, *Essay on the Theory of the Earth*, 90.

202 *"Like the prophet Ezekiel"*: Cohen, *Fate of the Mammoth*, 133.

202 *"A scrupulous examination" of their teeth and jaws*: Cuvier, *Essay on the Theory of the Earth*, 22.

202 *Cuvier was not done*: Ibid., 4.

202 *"All these facts . . . seem to me"*: Ibid., 24.

203 *Ninety-nine percent of all the species*: Coyne, *Why Evolution Is True*, 12.

203 *The history of life on Earth calls to mind*: These two sentences are a paraphrase of a remark of Carl Zimmer's in his superb *Evolution: The Triumph of an Idea*. See p. 171.

203 *"It doesn't seem so intelligent"*: Coyne, *Why Evolution Is True*, 12.

Chapter Thirty-one: Bursting the Limits of Time

205 *"the world previous to ours"*: Cuvier, *Essay on the Theory of the Earth*, 24.

206 *"I found myself as if placed in a charnel house"*: Wallace, *Beasts of Eden*, 8.

206 *(One of Cuvier's rivals)*: This was Lamarck, whose title was professor of "insects and worms."

206 *as the historian Martin Rudwick points out*: Rudwick, *Bursting the Limits of Time*, 368.

207 *"The bones being well-known"*: Ibid., 369.

207 *"Is not Cuvier the greatest poet"*: Honoré de Balzac, *The Wild Ass's Skin* (1831; repr., Oxford, UK: Oxford University Press, 2012), 19.

207 *"an antiquary of a new order"*: Göran Blix, *From Paris to Pompeii: French Romanticism and the Cultural Politics of Archaeology* (Philadelphia: University of Pennsylvania Press, 2009), 58.

207 *His aim, like that of conventional antiquarians*: Rudwick, *Earth's Deep History*, 150.

208 *"Would it not also be glorious"*: Cuvier, *Essay on the Theory of the Earth*, 13.

208 *"a man of realities"*: The passage is from chapter 2 of *Hard Times*, "Murdering the Innocents."

208 *"perhaps the finest intellect"*: Gould, *Time's Arrow*, 113.

Chapter Thirty-two: Boiling Seas and Exploding Mountains

209 *"None of the separate parts can change"*: Cuvier, *Essay on the Theory of the Earth*, 90.

210 *Cuvier had been the first*: Philippe Taquet and Kevin Padian, "The Earliest Known Restoration of a Pterosaur and the Philosophical Origins of Cuvier's *Ossemens Fossiles*," *Comptes Rendus Palevol* 3, no. 2 (March 2004).

210 *"Life, therefore, has been often disturbed"*: Cuvier, *Essay on the Theory of the Earth*, 16.

210 *"The march of nature is changed"*: Ibid., 24.

211 *Cuvier insisted that this apparent twist*: Gould, "The Stink Stones of Oeningen," in *Hen's Teeth and Horse's Toes*, 104.

211 *"These convulsions took on terrific"*: Worster, *Nature's Economy*, 138.

212 *Lamarck provided similar explanations*: Burkhardt, "Lamarck."

213 *"continents would crawl like amoebae"*: John McPhee, *In Suspect Terrain*, in McPhee, *Annals of the Former World*, 42.

213 *"No rose can ever remember"*: Cohen, *Fate of the Mammoth*, 240.

214 *"A universal upheaval, a catastrophe"*: J. B. Lamarck, *Système des animaux sans vertèbres* (Paris: Deterville, 1801). The passage also appears, in a slightly different translation, in Edey and Johanson, *Blueprints*, 29.

214 *"laboriously constructed vast edifices"*: Cuvier, "Elegy of Lamarck."

214 *"Only time and circumstance are needed"*: Ibid.

Chapter Thirty-three: Mayflies and Human History

215 *"creating and obliterating one form"*: Trevelyan, *A Shortened History of England* (London: Penguin, 1942), 446. I encountered Trevelyan's remark in Donald Worster's *Nature's Economy*, p. 13.

216 *"science every day more and more"*: Edward FitzGerald, *The Letters of Edward Fitz-Gerald*, vol. 1, *1830–1850* (Princeton, NJ: Princeton University Press, 1980), 476.

216 *filled "the human Soul with Wonder and Awe"*: Lescaze, *Paleoart*, 19.

216 *(Hindus believed that a single day)*: Alan Lightman, *Probable Impossibilities: Musings on Beginnings and Endings* (New York: Pantheon, 2021), 157.

216 *("the House of Millions of Years")*: Halliday, *Otherlands*, xiii.

216 *"Whether something is obvious"*: Christopher Ricks, *Along Heroic Lines* (New York: Oxford University Press, 2021), 139.

217 *If you took all the salt in the sea*: US Geological Survey, "Why Is The Ocean Salty?," May 23, 2019. Online at https://tinyurl.com/2kfu4at3.

217 *They dismissed the stars*: Hoskin, *Discoverers of the Universe*, 17.

217 *Herschel spent countless nights*: J. A. Bennett, "'On the Power of Penetrating into Space': The Telescopes of William Herschel," *Journal for the History of Astronomy* 7 (1976): 82.

218 *The poet Thomas Campbell*: I read about Thomas Campbell in Michael Hoskin's joint biography *Discoverers of the Universe: William and Caroline Herschel*, p. 179. Campbell's account of his conversation with Herschel can be found in his *Life and Letters of Thomas Campbell*, vol. 2 (New York: Harper, 1855), 13.

218 *(Galileo had tried, in 1638)*: Galileo tells the story in his *Dialogue Concerning Two Sciences*, trans. Henry Crew and Alfonso de Salvio (New York: The Macmillan Company, 1914). The book is written in dialogue form, and the passage appears in "Day One," p. 43. Online at https://tinyurl.com/mryzr9sk.

218 *"I have looked further into space"*: Hoskin, *Discoverers of the Universe*, 130.

218 *He felt "elevated and overcome"*: Ibid., 139.

218 *"It will strike terror into your soul"*: Balzac, *Wild Ass's Skin*, 19.

219 *it made human history "shrink into insignificance"*: Zimmerman, *Excavating Victorians*, 1.

219 *He did not downplay*: Gould, "Lyell's Pillars of Wisdom," in *Lying Stones of Marrakech*, 156.

219 *Lyell, whose religious views*: Bartholomew, "Lyell and Evolution," 266–67.

Chapter Thirty-four: Scattered by Desert Winds

221 *"an infant crying in the night"*: Tennyson, "In Memoriam AHH," canto 55.

221 *"To thousands, this is a sort of sacred book"*: Kirstie Blair, "Touching Hearts: Queen Victoria and the Curative Powers of *In Memoriam*," *Tennyson Research Bulletin* 7, no. 5 (November 2001): 246.

221 *(Queen Victoria herself copied)*: Victoria wrote out passages from the poem, including "'Tis better to have loved and lost." Online at https://tinyurl.com/mryawmah.

222 *"So careful of the type"*: Tennyson, "In Memoriam AHH," canto 56.

222 *"'So careful of the type?' but no"*: Ibid., canto 57.

223 *"Nature, red in tooth and claw"*: Ibid., canto 57.

223 *"blown about the desert dust"*: Ibid.

223 *"You speak of the flimsiness"*: Wedderburn and Cook, eds., *Works of John Ruskin*, 115.

223 *Arnold disliked and distrusted science*: Dudley, "Matthew Arnold and Science," especially pp. 276–81 and 292.

224 *"The Sea of Faith"*: Matthew Arnold, "Dover Beach" (1867).

225 *Victorians "went on loving their gardens"*: Worster, *Nature's Economy*, 127.

Chapter Thirty-five: Lizards in Scripture?

227 *"From the idea that one lost link"*: Eiseley, *Firmament of Time*, 48.

227 *"the surface of our globe has undergone"*: Cuvier, *Essay on the Theory of the Earth*, 239.

228 *That was why passages in the Bible*: Roberts, "Adam Sedgwick," 159.

229 *"It stood to reason that there was no mention"*: Chapman, *Caves, Coprolites and Catastrophes*, 88.

229 *a boa constrictor wrapping its "terrible coil"*: Miller, *Footprints of the Creator*, 320.

229 *"life is, in the main, enjoyment"*: Ibid.

230 *"The sheer scale and unanticipated strangeness"*: Rudwick, *Earth's Deep History*, 163.

Chapter Thirty-six: Dr. Jekyll and Mr. Hyde

231 *"As an anatomist"*: Flannery, "Dinosaur Crazy."

231 *"his lank forefinger followed up every line"*: Dickens, *David Copperfield*, chapter 16.

231 *a "clammy, irresponsive hand"*: Rupke, *Richard Owen*, 7.

232 *"As if by sympathetic magic"*: Ibid., 4.

232 *"The truth is," Thomas Huxley observed*: Ibid., 6.

232 *Jane Carlyle, the wittily malicious wife*: Ashton, *Thomas and Jane Carlyle*, 235.

233 *what he called an "anatomical passion"*: Owen, *Life of Richard Owen*, vol. I, 22.

233 *"pale, cold features and glassy, staring eyeballs"*: Ibid., 15.

233 *For the rest of his life he would retell*: Ibid., 21.

234 *Owen seemed "as if he was constantly attended"*: David Rains Wallace quotes the observation (from a naturalist named Edward Forbes) and makes the Jekyll and Hyde comparison in his *Beasts of Eden*, 19.

Chapter Thirty-seven: Defunct Animals and Open Windows

235 *the tattooed arms of South Sea islanders*: Semonin, *American Monster*, 249.

236 *Supposedly stood an amazing eight foot two*: Gina Kolata, "In a Giant's Story, a New Chapter Writ by His DNA," *New York Times*, January 5, 2011.

236 *the Royal Society arranged a private viewing*: Moore, *Knife Man*, 202.

236 *Trying to organize the fourteen thousand specimens*: Ibid., 237.

237 *"He published prolifically"*: Desmond, *Archetypes and Ancestors*, 22.

238 *a postmortem on a kangaroo*: Owen, *Life of Richard Owen*, vol. 1, 103.

238 *("Dickens enjoyed it like a schoolboy")*: Ibid., 302.

238 *"at best stupefyingly dull"*: Barber, *Heyday*, 176.

238 *("I determined I would never love")*: Horace Pym, ed., *Memories of Old Friends: Being Extracts from the Letters and Journals of Caroline Fox from 1835 to 1871* (London: Smith, Elder, 1883), 170.

238 *(Museum curators today pass along a story)*: Ashby, *Animal Kingdom*, 12. Ashby goes into a bit more detail about hoaxes in his *Platypus Matters: The Extraordinary Story of Australian Mammals* (Chicago: University of Chicago Press, 2022), 108.

238 *Now for a tasting!*: Ashby, *Animal Kingdom*, 12.

238 *To his delight, "minute drops"*: Owen, "Ornithorhynchus paradoxus," 523.

239 *"Today R. cut up the giraffe"*: Owen, *Life of Richard Owen*, vol. 1, 121.

239 *"The defunct rhinoceros arrived"*: Ibid.

240 *A "defunct elephant" was even worse*: Ibid., 296.

240 *"I made two ink outlines"*: Ibid., 104.

240 *"Engaged all day in drawing a wombat's brain"*: Owen, *Life of Richard Owen*, vol. 1, 101.

240 *"R. had a very bad night"*: Ibid., 295.

Chapter Thirty-eight: The Mystery of the Moa

241 *William was "to try his best"*: Frances Austin and Bernard Jones, "William Home Clift: The First Assistant Curator of the Hunterian Museum," *Annals of the Royal College of Surgeons of England* 62 (1980): 299.

242 *"naturally a great grief to Mr. Clift"*: Owen, *Life of Richard Owen*, vol. 1, 68.

242 *Caroline's mother had refused to allow*: Ibid., 35.

242 *In 1839 a sailor*: Ibid., 144.

242 *This was particularly bold*: Ibid., 246.

242 *Refusing to print his paper*: Ibid., 149.

243 *"He took, in our presence"*: Ibid., 151.

244 *"and in another minute there suddenly stalked"*: Ibid., 222.

Chapter Thirty-nine: "The Invention of Dinosaurs"

246 *"the suspicion of being a transmutationist"*: McGowan, *Dragon Seekers*, 163.

247 *"Disraeli was a self-made man"*: The remark was often attributed to the politician John Bright. See https://tinyurl.com/3be36u6u.

247 *"First there was nothing"*: The passage is from chapter 15 of *Tancred*.

247 *as a few daring thinkers had suggested*: I have in mind writers like Lord Monboddo and Erasmus Darwin. See Oscar Sherwin, "A Man with a Tail—Lord Monboddo," *Journal of the History of Medicine and Allied Sciences* 13, no. 4 (October 1958).

247 *"religion is a lie"*: John Willis Clark and Thomas McKenny Hughes, eds., *The Life and Letters of Adam Sedgwick*, vol. 2 (Cambridge, UK: Cambridge University Press, 1890), 84. The passage is from a letter to the geologist Charles Lyell written on April 9, 1845.

247 *the title of "the English Cuvier"*: Rupke, *Richard Owen*, 77.

248 *a report on "fossil reptiles of Great Britain"*: Torrens, "Politics and Paleontology," 178.

248 *"the greatest anatomist living"*: Ibid., 179.

248 *In 1842, in the published version of his lecture*: Historians and scientists often write that Owen coined the word *dinosaur* in 1841, but Hugh Torrens demonstrated in convincing detail that the correct date was 1842. Owen delivered his dinosaur talk in 1841, but Torrens found that he had *not* referred to dinosaurs in that oral presentation. Owen introduced the term in 1842, when his report was published in its full, written version. See Torrens, "Invention of Dinosaurs."

248 *in the admiring words of Stephen Jay Gould*: Gould, "An Awful Terrible Dinosaurian Irony," in *Lying Stones of Marrakech*, 191.

249 *"altogether peculiar among Reptiles"*: Owen, "Report on British Fossil Reptiles," 103.

250 *"I would propose the name of Dinosauria"*: Ibid., 145.

250 *(Owen missed half a dozen)*: Dean, *Mantell*, 190.

Chapter Forty: "When Troubles Come,
They Come Not Single Spies but in Battalions"

254 *("Richard Owen was the greatest anatomist")*: Gould, "A Seahorse for All Races," in *Leonardo's Mountain*, 124.

254 *"Owen did not see that others paved"*: Padian, "Quadrophenia," lxvii.

254 *"a clearly complex and unhappy character"*: Torrens and Cooper, "Uncurated Curators," 265.

254 *"I came home to my desolate hearth"*: Spokes, *Gideon Algernon Mantell*, 206.

254 *"stayed up very late reading"*: Owen, *Life of Richard Owen*, vol. 1, 212.

255 *friends had persuaded him*: Spokes, *Gideon Algernon Mantell*, 59.

255 *Visitors could purchase annual subscriptions*: Dean, *Gideon Mantell*, 152.

255 *Gideon could live in a bedroom*: Cadbury, *Terrible Lizard*, 212.

255 *"I have no companion"*: Spokes, *Gideon Algernon Mantell*, 118.

256 *("A large collection of fossil fresh-water shells")*: Ibid.

256 *The line stretched ninety wagons long*: Cadbury, *Terrible Lizard*, 223.

256 *so recently the property of Gideon Mantell*: Torrens, "Invention of Dinosaurs," 119.

256 *"Next day we had a geological outing"*: Owen, *Life of Richard Owen*, vol. 1, 166.

257 *"all [Mantell] had done was to collect fossils"*: Torrens, "Invention of Dinosaurs," 185.

257 *"all science is either physics"*: See Quote Investigator, "All Science Is Either Physics or Stamp Collecting," QuoteInvestigator.com, May 8, 2015, https://tinyurl .com/4dunh9f6.

257 *"I have again to regret a want of honour"*: Spokes, *Gideon Algernon Mantell*, 135.

Chapter Forty-one: Return of the Happy World

259 *Giddy writers competed to describe*: Lescaze, *Paleoart*, introduction by Walton Ford, 12. Ford cites several inventions and quotes Isaac Asimov's remark that "it was suddenly possible to ask, 'what would the future be like?' and expect a rational answer."

259 *The LocoMotive raced down the track*: Esposito, *A World History*, 150.

260 *"Even in those distant eras"*: Buckland, *Geology and Mineralogy*, 233.

261 *"There was no gradation or passage"*: Torrens, "Invention of Dinosaurs," 180.

261 *"the continuous operation of the ordained becoming"*: Desmond, *Archetypes and Ancestors*, 63.

261 *"It is obvious that it is the first duty"*: Thomas Huxley, *Man's Place in Nature*, 126. Darwin wrote Huxley a letter congratulating him on *Man's Place in Nature*. The "backwards, or forwards, or sideways" line, Darwin wrote, was "a delicious sneer, as good as a dessert." See https://tinyurl.com/bexbp6u8.

261 *"To what natural laws or secondary causes"*: Owen, *On the Nature of Limbs*, 22. Cohen discusses this passage in *Fate of the Mammoth*, 129.

262 *"Generations do not vary"*: Owen, *Anatomy of Vertebrates*, 808.

262 *"Nature has advanced with slow and stately steps"*: Owen, *On the Nature of Limbs*, 22.

262 *Archetypes were "Divine ideas"*: Owen, *Life of Richard Owen*, vol. 1, 388.

262 *"sustained, like an umbrella"*: Owen, *On the Nature of Limbs*, 7.

263 *animals differed in a crucial way*: Ibid., 9.

263 *"strange spiritual drama"*: Eiseley, *Firmament of Time*, 49.

264 *"The knowledge of such a being"*: Owen, *On the Nature of Limbs*, 85.

Chapter Forty-two: Dinner in a Dinosaur

265 *Only "lunatics" could have produced such "obscenities"*: Albert Wolff, "Le Calendrier Parisien," *Le Figaro*, April 3, 1876. Online at https://tinyurl.com/752vhzfe. The remarks about the Impressionists and their "obscenities" appear on p. 1, in the fourth column, under the heading "Dimanche 2" (i.e., Sunday, April 2, 1876).

265 *"a truly tragic street lamp"*: Hippolyte Blancard, "Untitled," April 1889. The quotation comes from a Museum of Modern Art webpage online at https://www.moma.org/collection/works/88152.

265 *"odious meowing"*: Nicolas Slonimsky, *Lexicon of Musical Invective: Critical Assaults on Composers Since Beethoven's Time* (New York: Norton, 2000), 9.

266 *"It is mean and miserable"*: Rev. Edward Henry Carr, *The Nation Admonished: A Sermon Preached (in Part) in Christ Chapel, Maida Hill, December 22, 1861, the Day Before the Funeral of the Late Prince Consort* (London: Wertheim, 1862), 6.

266 *Twenty-two of Britain's most prominent*: Rupke, *Richard Owen*, 79.

266 *He'd been commissioned to build*: Witton and Michel, *Art and Science*, 10.

267 *(The iguanodon alone stretched thirty-five feet)*: "Dinner to Professor Owen in the Iguanodon," *Leader*, January 7, 1854.

268 *"Crystal Palace. Mr. B. Waterhouse Hawkins requests"*: The card is reproduced in *Paleoart* by Zoë Lescaze, p. 64.

269 *Special trains had to be rushed*: Desmond, "Designing the Dinosaur," 228.

269 *Two million visitors a year*: Lescaze, *Paleoart*, 64.

270 *The sculptures occupied a landscape*: Marshall, "A Dim World," 295.

270 *(Hawkins took advantage of the lakes)*: Peck, "Art of Bones."

270 *"The sensational statues acted on viewers"*: Lescaze, *Paleoart*, 64.

271 *(the historian of science Nicolaas Rupke points out)*: Rupke, *Richard Owen*, 80.

271 *(Owen thought so, too)*: Ibid., 79.

271 *"an expression of successful conquest"*: Rudwick, *Deep Time*, 148.

271 *"Saurians, Pterodactyls all!"*: Cadbury, *Terrible Lizard*, 298.

272 *the Thames was choked with ice*: David B. Williams has written the best account of the Crystal Palace dinner, in his "Benchmarks: December 31, 1853: Dinner in a Dinosaur." Williams generously shared several finds that he had unearthed, including a newspaper's description of the weather in London on December 31, 1853.

Chapter Forty-three: "It Is Like Confessing a Murder"

273 *He was so protective of his privacy*: Desmond and Moore, *Darwin*, 306.

275 *"It is," he wrote, "like confessing a murder"*: Letter from Darwin to Joseph Hooker, January 11, 1844, University of Cambridge Darwin Correspondence Project. Online at https://tinyurl.com/nhhcb5vs.

275 *(it was Owen he enlisted)*: Browne, *Charles Darwin*, 348–51.

275 *"What differentiates revolutionary thinkers"*: Sulloway, *Born to Rebel*, 20.

275 *"the single best idea anyone has ever had"*: Dennett, *Darwin's Dangerous Idea*, 21.

276 *"IN ORDER TO MAKE A PERFECT AND BEAUTIFUL MACHINE"*: Ibid., 65.

276 *"What but the wolf's tooth"*: The lines are from "The Bloody Sire," written in 1940.

277 *Robber barons and magnates rushed*: Rieppel, "How Dinosaurs Became Tyrants,"
     777.
277 *"a dim world, where monsters dwell"*: Marshall, "A Dim World," 296.

Epilogue

279 *winners write the history books*: The remark, which has been attributed to both Winston Churchill and Hermann Göring (!), has a long history of its own. See Matthew Phelan, "The History of 'History Is Written by the Victors,'" *Slate*, November 26, 2019.
279 *"The peaceful career of this indefatigable cultivator"*: Jacob Gruber, "Sir Richard Owen," *Dictionary of National Biography* (Oxford, UK: Oxford University Press, 2004), September 23, 2004.
279 *"I used to be ashamed of hating him"*: Darwin complained about Owen in an August 4, 1872, letter to his friend Joseph Hooker. See https://tinyurl.com/4my9t447.
281 *"ill with symptoms of paralysis"*: Fairbank, "William Adams and the Spine of Gideon Mantell."
281 *"Could I choose my destiny"*: Ibid.
281 *"Tumour of considerable size"*: Ibid.
281 *"the severest degree of deformity"*: Ibid.
282 *(it was Mantell who had suggested)*: Spokes, *Gideon Algernon Mantell*, 327. Spokes's biography includes a section called "Dr. Mantell's Spine" that sets out the story.

Bibliography

Allen, David Elliston. *The Naturalist in Britain: A Social History*. Princeton, NJ: Princeton University Press, 1994.

Altholz, Josef. "The Warfare of Conscience with Theology." In Josef Altholz, ed., *The Mind and Art of Victorian England*. Minneapolis, MN: University of Minnesota Press, 1976.

Annan, Noel. *The Dons: Mentors, Eccentrics, and Geniuses*. Chicago: University of Chicago Press, 1999.

Aries, Roger. "Leibniz on the Unicorn and Various Other Curiosities." *Early Science and Medicine* 3, no. 4 (1998).

Armstrong, Carol, and Catherine de Zegher, eds. *Ocean Flowers: Impressions from Nature*. Princeton, NJ: Princeton University Press, 2004.

Ashby, Jack. *Animal Kingdom: A Natural History in 100 Objects*. Cheltenham, UK: History Press, 2017.

Ashton, Rosemary. *Thomas and Jane Carlyle: Portrait of a Marriage*. London: Random House UK, 2002.

Asma, Stephen T. *Stuffed Animals and Pickled Heads: The Culture and Evolution of Natural History Museums*. New York: Oxford University Press, 2001.

Barber, Lynn. *The Heyday of Natural History, 1820–1870*. Garden City, NY: Doubleday, 1980.

Barnett, Cynthia. *The Sound of the Sea: Seashells and the Fate of the Oceans*. New York: Norton, 2021.

Barrow, Mark. *Nature's Ghosts: Confronting Extinction from the Age of Jefferson to the Age of Ecology*. Chicago: University of Chicago Press, 2009.

Bartholomew, Michael. "Lyell and Evolution: An Account of Lyell's Response to the Prospect of an Evolutionary Ancestry for Man." *British Journal for the History of Science* 6, no. 3 (June 1973).

Bayne, Peter, ed. *The Life and Letters of Hugh Miller.* Vol. 1. London: Strahan, 1871.

Beecher, Henry Ward. *Evolution and Religion.* New York: Fords, Howard and Hulbert, 1885.

Benton, Michael. *Dinosaurs Rediscovered: The Scientific Revolution in Paleontology.* New York: Thames and Hudson, 2019.

Bigelow, Jacob. "Bigelow on the Sea Serpent." *American Journal of Science and Arts* 2 (1820).

Black, Riley. *Last Days of the Dinosaurs: An Asteroid, Extinction, and the Beginning of Our World.* New York: St. Martin's, 2022.

Blair, Kirstie. "Touching Hearts: Queen Victoria and the Curative Powers of *In Memoriam.*" *Tennyson Research Bulletin* 7, no. 5 (November 2001).

Blei, Daniela. "Inventing the Beach: The Unnatural History of a Natural Place." *Smithsonian*, June 23, 2016.

Bowler, Peter J. *Fossils and Progress.* New York: Science History, 1976.

Boyd, Julian P. "The Megalonyx, the Megatherium, and Thomas Jefferson's Lapse of Memory." *Proceedings of the American Philosophical Society* 102, no. 5 (October 20, 1958).

Boyle, Robert. *The Excellence of Theology.* (Originally published 1674.) Online at https://tinyurl.com/42huy2ry.

——. *A Disquisition About the Final Causes of Natural Things.* London: HC, 1688. Online at https://tinyurl.com/5vmpw8c2.

Broberg, Gunnar. *The Man Who Organized Nature: The Life of Linnaeus.* Princeton, NJ: Princeton University Press, 2023.

Brooke, John Hedley. *Science and Religion: Some Historical Perspectives.* New York: Cambridge University Press, 2014.

Browne, Janet. *Charles Darwin: A Biography.* Vol. 1, *Voyaging.* Princeton, NJ: Princeton University Press, 1996.

——. "Noah's Flood, the Ark, and the Shaping of Early Modern Natural History." In David Lindberg and Ronald Numbers, eds., *When Science and Christianity Meet.* Chicago: University of Chicago Press, 2003.

Buckland, William. "Notice on the Megalosaurus or Great Fossil Lizard of Stonesfield." *Transactions of the Geological Society of London*, series 2, vol. 1 (1824).

——. *Geology and Mineralogy Considered with Reference to Natural Theology.* Vol. 1. Treatise 6 of *The Bridgewater Treatises on the Power, Wisdom, and Goodness of God as Manifested in the Creation.* London: Pickering, 1836.

Burek, Cynthia V. "The First Female Fellows and the Status of Women in the Geological Society of London." Geological Society, London, Special Publications 317 (January 1, 2009).

Burkhardt Jr., William. "Lamarck, Evolution, and the Inheritance of Acquired Characteristics." *Genetics* 194, no. 4 (August 2013).

Cadbury, Deborah. *Terrible Lizard: The First Dinosaur Hunters and the Birth of a New Science.* New York: Henry Holt, 2001.

Campbell, Thomas. *Life and Letters of Thomas Campbell.* Vol. 2. London: Hall, Virtue, 1850.

Carroll, Sean. *Remarkable Creatures: Epic Adventures in the Search for the Origins of Species.* Boston: Mariner, 2009.

Casey, John. *After Lives: A Guide to Heaven, Hell, and Purgatory.* New York: Oxford University Press, 2009.

Chapman, Allan. *Caves, Coprolites and Catastrophes: The Story of Pioneering Geologist and Fossil-Hunter William Buckland.* London: SPCK, 2020.

Christie, Ann. "A Taste for Seaweed: William Kilburn's Late Eighteenth-Century Designs for Printed Cottons." *Journal of Design History* 24, no. 4 (2011).

Clark, John F. M. *Bugs and the Victorians.* New Haven, CT: Yale University Press, 2009.

Clark, John Willis, and Thomas McKenny, eds. *The Life and Letters of Adam Sedgwick.* Vol. 2. Cambridge, UK: Cambridge University Press, 1890.

Cohen, Claudine. *The Fate of the Mammoth: Fossils, Myth, and History.* Chicago: University of Chicago Press, 2002.

Cohn, Norman. *Noah's Flood: The Genesis Story in Western Thought.* New Haven, CT: Yale University Press, 2009.

Coleridge, Christabel Rose. *Charlotte Mary Yonge: Her Life and Letters.* New York: Macmillan, 1903.

Conniff, Richard. "Mammoths and Mastodons: All-American Monsters." *Smithsonian,* April 2010.

———. *The Species Seekers: Heroes, Fools, and the Mad Pursuit of Life on Earth.* New York: Norton, 2011.

Coonen, Lester P., and Charlotte M. Porter. "Thomas Jefferson and American Biology." *BioScience* 26, no. 12 (December 1976).

Corbin, Alain. *The Lure of the Sea: The Discovery of the Seaside in the Western World, 1750–1840.* New York: Penguin, 1995.

Coyne, Jerry. *Why Evolution Is True.* New York: Viking, 2009.

Cutler, Alan. *The Seashell on the Mountaintop: A Story of Science, Sainthood, and the Humble Genius Who Discovered a New History of the Earth.* New York: Dutton, 2003.

Cuvier, Georges. "Memoir on the Species of Elephants, Both Living and Fossil." April 4, 1796; reprinted in Martin Rudwick, ed., *Georges Cuvier, Fossil Bones, and Geological Catastrophes: New Translations and Interpretations of the Primary Texts.* Chicago: University of Chicago Press, 1998.

———. *Essay on the Theory of the Earth.* Edinburgh: Blackwood, 1817.

———. "Elegy of Lamarck." *Edinburgh New Philosophical Journal* 20 (January 1836).

Daston, Lorraine, and Katharine Park. *Wonders and the Order of Nature, 1150–1750.* New York: Zone Books, 1998.

Davis, Larry E. "Mary Anning of Lyme Regis: 19th Century Pioneer in British Paleontology." *Headwaters: The Faculty Journal of the College of Saint Benedict and Saint John's
   University* 26 (May 22, 2012).

Davis, Ted. "The Faith of a Great Scientist: Robert Boyle's Religious Life, Attitudes,
   and Vocation." *BioLogos*, August 8, 2013.

Dean, Dennis. *Gideon Mantell and the Discovery of Dinosaurs.* New York: Cambridge
   University Press, 1999.

Dennett, Daniel. *Darwin's Dangerous Idea: Evolution and the Meanings of Life.* New York:
   Simon and Schuster, 1996.

Desmond, Adrian. *The Hot-Blooded Dinosaurs: A Revolution in Paleontology.* New York:
   Dial Press, 1976.

———. "Designing the Dinosaur: Richard Owen's Response to Robert Edmont Grant."
   Isis 70, no. 2 (June 1979).

———. *Archetypes and Ancestors: Paleontology in Victorian London, 1850–1875.* Chicago:
   University of Chicago Press, 1984.

———. *The Politics of Evolution: Morphology, Medicine, and Reform in Radical London.*
   Chicago: University of Chicago Press, 1992.

Desmond, Adrian, and James Moore. *Darwin: The Life of a Tormented Evolutionist.* New
   York: Norton, 1994.

Dudley, Fred A. "Matthew Arnold and Science." *Proceedings of the Modern Language
   Association* 57, no. 1 (March 1942).

Dugatkin, Lee Alan. *Mr. Jefferson and the Giant Moose: Natural History in Early America.*
   Chicago: University of Chicago Press, 2009.

———. "Buffon, Jefferson and the Theory of New World Degeneracy." *Evolution: Education and Outreach* 12 (2019).

———. *Behind the Crimson Curtain: The Rise and Fall of Peale's Museum.* New York: Butler
   Books, 2020.

Dugatkin, Lee Alan, and Carl Bergstrom. *Evolution.* New York: Norton, 2016.

Edey, Maitland, and Donald Johanson. *Blueprints: Solving the Mystery of Evolution.* Boston: Little, Brown, 1989.

Eiseley, Loren. *The Immense Journey.* New York: Vintage, 1946.

———. *The Firmament of Time.* New York: Atheneum, 1960.

———. *Darwin's Century: Evolution and the Men Who Discovered It.* New York: Anchor,
   1961.

Ellis, Richard. *Sea-Dragons: Predators of the Prehistoric Oceans.* Lawrence, KS: University
   Press of Kansas, 2003.

Elson, Ruth Miller. *Guardians of Tradition: American Schoolbooks of the Nineteenth Century.*
   Lincoln, NE: University of Nebraska Press, 1964.

Emling, Shelley. *The Fossil Hunter: Dinosaurs, Evolution, and the Woman Whose Discoveries
   Changed the World.* New York: St. Martin's, 2009.

Endersby, Jim. "Creative Designs? How Darwin's *Origin* Caused the Victorian Crisis of Faith, and Other Myths." *Times Literary Supplement*, March 14, 2007.

Enenkel, Karl, and Paul Smith, eds. *Zoology in Early Modern Culture: Intersections of Science, Theology, Philology, and Political and Religious Education*. Leiden, Netherlands: Brill, 2014.

Esposito, Matthew, ed. *A World History of Railroad Culture, 1830–1930*. Vol. 1, *The United Kingdom*. London: Routledge, 2020.

Fairbank, Jeremy. "William Adams and the Spine of Gideon Algernon Mantell." *Annals of the Royal College of Surgeons of England* 86 (2004).

Fisch, Harold. "The Scientist as Priest: A Note on Robert Boyle's Natural Theology." *Isis* 4, no. 3 (September 1953).

FitzGerald, Edward. *The Letters of Edward FitzGerald*. Vol. 1, *1830–1850*. Princeton, NJ: Princeton University Press, 2017.

Flannery, Tim. "Dinosaur Crazy." *New York Review of Books*, January 17, 2002.

Fraser, Antonia. *Perilous Question: Reform or Revolution? Britain on the Brink, 1832*. New York: PublicAffairs, 2013.

Freeman, Michael. *Victorians and the Prehistoric: Tracks to a Lost World*. New Haven, CT: Yale University Press, 2004.

Gale, Barry. "Darwin and the Concept of a Struggle for Existence: A Study in the Extra-scientific Origins of Scientific Ideas." *Isis* 63, no. 3 (September 1972).

Gates, Barbara T. "Introduction: Why Victorian Natural History." *Victorian Culture and Literature* 35, no. 2 (2007).

Geikie, Archibald. *Life of Sir Roderick Murchison*. Vol. 1. London: Murray, 1875.

Gerbi, Antonello. *The Dispute of the New World: The History of a Polemic, 1750–1900*. Pittsburgh: University of Pittsburgh, 1973.

Gilette, David L., and Martin G. Lockley, eds. *Dinosaur Tracks and Traces*. New York: Cambridge University Press, 1989.

Gillispie, Charles C. *Genesis and Geology: A Study in the Relations of Scientific Thought, Natural Theology, and Social Opinion in Great Britain*. Cambridge, MA: Harvard University Press, 1951.

Glassie, John. *A Man of Misconceptions: The Life of an Eccentric in an Age of Change*. New York: Riverhead, 2012.

Gleick, James. *Time Travel: A History*. New York: Vintage, 2016.

Glendening, John. "'The World-Renowned Ichthyosaurus': A Nineteenth-Century Problematic and Its Representations." *Journal of Literature and Science* 2, no. 1 (2009).

Gliserman, Susan. "Early Victorian Science Writers and Tennyson's 'In Memoriam': A Study in Cultural Exchange: Part I." *Victorian Studies* 18, no. 3 (March 1975).

Goodhue, Thomas. "The Faith of a Fossilist: Mary Anning." *Anglican and Episcopal History* 70, no. 1 (March 2001).

———. "Mary Anning: The Fossilist as Exegete." *Endeavour* 29, no. 1 (March 2005).

———. "Origins of Paleontology and the Impact of Religion on the Development of Evolutionary Theory." *Scientific American*, February 13, 2009. (This is the transcript of a podcast featuring Goodhue.)

Gooding, Francis. "Feathered, Furred or Coloured: The Dying of the Dinosaurs." *London Review of Books*, February 22, 2018. (This is a review of *Paleoart: Visions of the Prehistoric Past* by Zoë Lescaze.)

Goodman, Ruth. *How to Be a Victorian: A Dawn-to-Dusk Guide to Victorian Life.* New York: Norton, 2013.

Gopnik, Alison. "How Animals Think." *Atlantic*, May 2016. (This is a review of *Are We Smart Enough to Know How Smart Animals Are?* by Frans de Waal.)

Gordon, Elizabeth Oke. *The Life and Correspondence of William Buckland.* London: Murray, 1894.

Gosse, Edmund. *Father and Son: A Study of Two Temperaments.* New York: Scribner's, 1907.

Gosse, Philip. *Omphalos: An Attempt to Untie the Geological Knot.* London: Van Voorst, 1857.

Gould, Stephen Jay. *Hen's Teeth and Horse's Toes: Further Reflections in Natural History.* New York: North, 1983.

———. *The Flamingo's Smile: Reflections in Natural History.* New York: Norton, 1985.

———. *Time's Arrow, Time's Cycle: Myth and Metaphor in the Discovery of Geological Time.* Cambridge, MA: Harvard University Press, 1987.

———. *Eight Little Piggies: Reflections in Natural History.* New York: Norton, 1993.

———. *Dinosaur in a Haystack: Reflections in Natural History.* New York: Harmony, 1995.

———. *Leonardo's Mountain of Clams and the Diet of Worms.* New York: Three Rivers Press, 1998.

———. "The Man Who Invented Natural History." *New York Review of Books*, October 22, 1998.

———. *The Lying Stones of Marrakech: Penultimate Reflections in Natural History.* New York: Harmony, 2000.

Greenblatt, Stephen. *Marvelous Possessions: The Wonder of the New World.* Chicago: University of Chicago Press, 1991.

———. *Swerve: How the World Became Modern.* New York: Norton, 2012.

Greene, John. *The Death of Adam: Evolution and Its Impact on Western Thought.* Ames, IA: Iowa State University Press, 1959.

Greene, Mott. "Genesis and Geology Revisited: The Order of Nature and the Nature of Order in Nineteenth Century Britain." In David Lindberg and Ronald Numbers, eds., *When Science and Christianity Meet.* Chicago: University of Chicago Press, 2003.

Gregory, James. "Eccentric Biography and the Victorians." *Biography* 30, no. 3 (Summer 2007).

Grigson, Caroline. *Menagerie: The History of Exotic Animals in England.* New York: Oxford University Press, 2016.

Gruber, Jacob W., and John G. Thackray. *Richard Owen and His Correspondents.* London: Natural History Museum Publications, 1992.

Halley, Edmund. "A Short Account of the Cause of the Saltness of the Ocean, and . . . the Age of the World." *Philosophical Transactions of the Royal Society*, August 31, 1715.

Halliday, Thomas. *Otherlands: A Journey Through Earth's Extinct Worlds*. New York: Random House, 2022.

Hare, Augustus. *Story of My Life*. Vol. 5. London: George Allen, 1900.

Hawkins, Thomas. *Memoirs of Ichthyosauri and Plesiosauri, Extinct Monsters of the Ancient Earth*. London: Rolfe and Fletcher, 1834.

Hendrikx, Sophia. "Monstrosities from the Sea: Taxonomy and Tradition in Conrad Gessner's (1516–1565) Discussion of Cetaceans and Sea Monsters." *Anthropozoologica* 53, no. 11 (2018).

Himmelfarb, Gertrude. *Marriage and Morals Among the Victorians and Other Essays*. New York: Vintage, 1987.

Hitchcock, Edward. "Description of Two New Species of Fossil Footprints Found in Massachusetts and Connecticut, or, of the Animals That Made Them." *American Journal of Science* (1847).

———. *Ichnology of New England*. Boston: William White, 1858.

Hoare, Philip. *Albert and the Whale: Albrecht Dürer and How Art Imagines Our World*. New York: Pegasus, 2021.

Hooke, Robert. *Micrographia*. London: Royal Society, 1665.

———. *The History and Philosophy of Earthquakes*. Cambridge, UK: Cambridge University Press, 2013. (Originally published in London in 1767.)

Hoskin, Michael. *William Herschel and the Construction of the Heavens*. New York: Norton, 1964.

———. *Discoverers of the Universe: William and Caroline Herschel*. Princeton, NJ: Princeton University Press, 2011.

Hughes-Hallett, Penelope. *The Immortal Dinner: A Famous Evening of Genius and Laughter in Literary London, 1817*. New York: New Amsterdam Books, 2002.

Hunt, Stephen E. "'Free, Bold, Joyous': The Love of Seaweed in Margaret Gatty and Other Mid-Victorian Writers." *Environment and History* 11, no. 1 (February 2005).

Huxley, Aldous. *Wordsworth in the Tropics*. London: Yale University Press, 1928.

Huxley, Thomas. *Man's Place in Nature*. 1863. Reprint, New York: Modern Library, 2001.

———. "On a Piece of Chalk." *Macmillan's Magazine*, 1868. Online at https://tinyurl.com/4mwtbz5x.

Isaacson, Walter. *Leonardo da Vinci*. New York: Simon and Schuster, 2018.

Jardine, Lisa. *The Curious Life of Robert Hooke: The Man Who Measured London*. New York: Harper, 2004.

Jefferson, Thomas. *Notes on the State of Virginia*. 1785. Reprint, Richmond, VA: Randolph, 1853.

———. "A Memoir on the Discovery of Certain Bones of a Quadruped of the Clawed Kind in the Western Parts of Virginia." *Transactions of the American Philosophical Society* 4 (1799).

Kenyon, T. K. "Science and Celebrity: Humphry Davy's Rising Star." *Distillations*, December 22, 2008.

Kolbert, Elizabeth. "The Lost World." *New Yorker*, December 16, 2013.

Le Guyader, Hervé. *Geoffroy Saint-Hilaire: A Visionary Naturalist*. Chicago: University of Chicago Press, 2004.

Lescaze, Zoë. *Paleoart: Visions of the Prehistoric Past*. Cologne, Germany: Taschen, 2017.

Logan, Thad. *The Victorian Parlour: A Cultural Study*. New York: Cambridge University Press, 2004.

Lovejoy, Arthur. *The Great Chain of Being: A Study of the History of an Idea*. Cambridge, MA: Harvard University Press, 1976.

Lurie, Edward. *Louis Agassiz: A Life in Science*. Chicago: University of Chicago Press, 1960.

Lyell, K. S., ed. *Life, Letters and Journals of Sir Charles Lyell, Bart*. Vol. 1. London: Murray, 1881.

Mantell, Gideon. *Illustrations of the Geology of Sussex*. London: Lupton Relfe, 1827.

———. "The Geological Age of Reptiles." *Edinburgh New Philosophical Journal* 11 (April–September 1831).

———. *The Geology of the South-east of England*. London: Longman, 1833.

———. *The Wonders of Geology*. London: Henry Bohn, 1848.

———. *The Medals of Creation*. London: Henry Bohn, 1855.

Marshall, Nancy Rose. "A Dim World, Where Monsters Dwell: The Spatial Time of the Sydenham Crystal Palace Dinosaur Park." *Victorian Studies* 49, no. 2 (Winter 2007).

Mayor, Adrienne. *The First Fossil Hunters: Dinosaurs, Mammoths, and Myths in Greek and Roman Times*. Princeton, NJ: Princeton University Press, 2011.

———. *Flying Snakes and Griffin Claws: And Other Classical Myths, Historic Oddities, and Scientific Curiosities*. Princeton, NJ: Princeton University Press, 2022.

McGowan, Christopher. *The Dragon Seekers: How an Extraordinary Circle of Fossilists Discovered the Dinosaurs and Paved the Way for Darwin*. New York: Basic Books, 2001.

McKay, John. *Discovering the Mammoth: A Tale of Giants, Unicorns, Ivory, and the Birth of a New Science*. New York: Pegasus, 2017.

McPhee, John. *Annals of the Former World*. New York: Farrar, Straus and Giroux, 2000.

Meadows, A. J. "Astronomy and Geology, Terrible Muses! Tennyson and 19th-Century Science." *Notes and Records of the Royal Society of London* 46, no. 1 (January 1992).

Medawar, P. B., and J. S. Medawar. *Aristotle to Zoos: A Philosophical Dictionary of Biology*. London: Weidenfeld and Nicolson, 1984.

Miller, Hugh. *The Footprints of the Creator*. Boston: Gould and Lincoln, 1850.

———. *The Testimony of the Rocks*. Boston: Gould and Lincoln, 1857.

Mitchell, Sally. *Daily Life in Victorian England*. Westport, CT: Greenwood, 1996.

Moore, John. *Science as a Way of Knowing: The Foundations of Modern Biology*. Cambridge, MA: Harvard University Press, 1993.

Moore, Wendy. *The Knife Man: Blood, Body Snatching, and the Birth of Modern Surgery.* New York: Broadway, 2005.

Morris, Amy. "Geomythology on the Colonial Frontier: Edward Taylor, Cotton Mather, and the Claverack Giant." *William and Mary Quarterly* 70, no. 4 (October 2013).

Morrison, Robert. *The Regency Years: During Which Jane Austen Writes, Napoleon Fights, Byron Makes Love, and Britain Becomes Modern.* New York: Norton, 2020.

Natural History Museum. *Nature's Cathedral: A Celebration of the Natural History Museum Building.* London: Natural History Museum, 2020.

O'Connor, Ralph. *The Earth on Show: Fossils and the Poetics of Popular Science, 1802–1856.* Chicago: University of Chicago Press, 2007.

Ospovat, Dov. *The Development of Darwin's Theory: Natural History, Natural Theology, and Natural Selection, 1838–1859.* New York: Cambridge University Press, 1981.

Owen, Richard. "On the Mammary Glands of the Ornithorhynchus paradoxus." *Philosophical Transactions of the Royal Society of London* 122 (1832).

———. "Report on British Fossil Reptiles." In *Report of the Eleventh Meeting of the British Association for the Advancement of Science, Held at Plymouth in July 1841.* London: John Murray, 1842.

———. *On the Nature of Limbs.* 1849. Reprint, Chicago: University of Chicago Press, 2007.

———. *On the Anatomy of Vertebrates.* Vol. 3. London: Longmans, Green, 1868.

Owen, Richard. *The Life of Richard Owen, by His Grandson.* London: Murray, 1894.

Padian, Kevin. "Richard Owen's Quadrophenia: The Pull of Opposing Forces in Victorian Cosmogony." In Richard Owen, *On the Nature of Limbs,* ed. Ron Amundsen. Chicago: University of Chicago Press, 2007.

Paley, William. *Natural Theology.* London: Faulder, 1802.

Pantin, C. F. A. "Alfred Russel Wallace, F.R.S., and His Essays of 1858 and 1855." *Notes and Records of the Royal Society of London* 14, no. 1 (June 1959).

Peck, Robert McCracken. "The Art of Bones." *Natural History,* December 2008–January 2009.

Pike, E. Royston. *Golden Times: Human Documents of the Victorian Age.* New York: Schocken, 1972.

Pinto-Correia, Clara. *Return of the Crazy Bird: The Sad, Strange Tale of the Dodo.* New York: Springer-Verlag, 2003.

Plot, Robert. *The Natural History of Oxfordshire.* London: Lichfield, 1705.

Plumb, Christopher. "'The Queen's Ass': The Cultural Life of Queen Charlotte's Zebra in Georgian England." In Samuel J. Alberti, ed., *The Afterlives of Animals: A Museum Menagerie.* Charlottesville, VA: University of Virginia Press, 2011.

Plumb, J. H. *The First Four Georges.* London: Batsford, 1956.

Plumly, Stanley. *The Immortal Evening: A Legendary Dinner with Keats, Wordsworth, and Lamb.* New York: Norton, 2014.

Poskett, James. *Horizons: The Global Origins of Modern Science.* Boston: Mariner, 2022.

Preston, Douglas. "The Day the Dinosaurs Died." *New Yorker,* March 29, 2019.

Prothero, Donald R. *The Story of the Dinosaurs in 25 Discoveries: Amazing Fossils and the People Who Found Them.* New York: Columbia University Press, 2019.

Prothero, Donald R., and Robert M. Schoch. *Horns, Tusks, and Flippers: The Evolution of Hoofed Animals.* Baltimore: Johns Hopkins University Press, 2002.

Pym, Horace, ed. *Memories of Old Friends, Being Extracts from the Letters and Journals of Caroline Fox from 1835 to 1871.* London: Smith, Elder, 1882.

Rappaport, Rhoda. "Hooke on Earthquakes: Lectures, Strategy, and Audience." *British Journal for the History of Science* 19, no. 2 (July 1986).

Ray, John. *Three Physico-theological Discourses.* London: Innys, 1713.

Richardson, George. *Sketches in Prose and Verse, Containing Visits to the Mantellian Museum.* London: Relfe and Fletcher, 1838.

Ricks, Christopher, ed. *Tennyson: A Selected Edition.* Berkeley, CA: University of California Press, 1989.

Ridley, Mark. "Dreadful Beasts." *London Review of Books,* June 28, 1990.

Rieppel, Lukas. "How Dinosaurs Became Tyrants of the Prehistoric." *Environmental History* 25 (October 2020).

Ritchie, Robert C. *The Lure of the Beach: A Global History.* Berkeley, CA: University of California Press, 2021.

Roberts, Michael B. "Adam Sedgwick (1785–1873): Geologist and Evangelical." *Geological Society, London, Special Publications* 310 (March 2009).

Roger, Jacques. *Buffon: A Life in Natural History.* Ithaca, NY: Cornell University Press, 1997.

Rudwick, Martin. *Scenes from Deep Time: Early Pictorial Representations of the Prehistoric World.* Chicago: University of Chicago Press, 1992.

———. *Bursting the Limits of Time: The Reconstruction of Geohistory in the Age of Revolution.* Chicago: University of Chicago Press, 2007.

———. *Earth's Deep History: How It Was Discovered and Why It Matters.* Chicago: University of Chicago Press, 2014.

Rupke, Nicolaas. *The Great Chain of History: William Buckland and the English School of Geology, 1814–1849.* New York: Oxford University Press, 1983.

———. *Richard Owen: Biology Without Darwin.* Chicago: University of Chicago Press, 2009.

Russell, Richard. *A Dissertation on the Use of Seawater in the Diseases of the Glands.* London: Owen, 1769.

Sage, Victor. "Dickens and Professor Owen: Portrait of a Friendship." *Sillages Critiques* 2 (2001).

Schiermeier, Quirin. "The Megaflood That Made Britain an Island." *Nature,* July 18, 2007.

Schiff, Stacy. *The Witches: Suspicion, Betrayal, and Hysteria in 1692 Salem*. Boston: Back Bay, 2016.

Secord, James. *Victorian Sensation: The Extraordinary Publication, Reception, and Secret Authorship of "Vestiges of the Natural Creation."* Chicago: University of Chicago Press, 2003.

Semonin, Paul. *American Monster: How the Nation's First Prehistoric Creature Became a Symbol of National Identity*. New York: New York University Press, 2000.

———. "Peale's Mastodon: The Skeleton in Our Closet." *Commonplace*, January 2004.

Sharpe, Tom. *The Fossil Woman: A Life of Mary Anning*. Wimborne, UK: Dovecote, 2021.

Shortland, Michael. "Darkness Visible: Underground Culture in the Golden Age of Geology." *British Journal for the History of Science* 32 (March 1, 1994).

Simpson, George Gaylord. "The Beginnings of Vertebrate Paleontology in North America." *Proceedings of the American Philosophical Society* 86, no. 1 (September 25, 1942).

Spokes, Sidney. *Gideon Algernon Mantell, Surgeon and Geologist*. London: Bale, 1927.

Stanford, Donald E. "The Giant Bones of Claverack, New York, 1705." *New York History* 40, no. 1 (January 1959).

Stott, Rebecca. *Darwin's Ghosts: The Secret History of Evolution*. New York: Spiegel and Grau, 2014.

Sulloway, Frank. *Born to Rebel: Birth Order, Family Dynamics, and Creative Lives*. New York: Vintage, 1997.

———. "Why Darwin Rejected Intelligent Design." *Journal of Biosciences* 34, no. 2 (June 2009).

Tattersall, Ian. *The Strange Case of the Rickety Cossack: And Other Cautionary Tales from Human Evolution*. New York: St. Martin's, 2015.

Taylor, Michael A., and Lyall I. Anderson. "Tennyson and the Geologists Part 2: Saurians and the Isle of Wight." *Tennyson Research Bulletin* 10, no. 5 (November 2016).

Tennyson, Alfred, Lord. "In Memoriam AHH." In Alfred Lord Tennyson, *Poems*. London: Macmillan, 1908. Online at https://tinyurl.com/yrdcy328. (Originally published in 1850.)

Thompson, Edward P. *The Note-book of a Naturalist*. London: Smith, Elder, 1845.

Thomson, Keith. "Jefferson's Old Bones." *American Scientist*, May–June 2011.

Too, Kathryn S. "Strangers in the House of God: Cotton Mather, Onesimus, and an Experiment in Christian Slaveholding." *Proceedings of the American Antiquarian Society* 117 (2007).

Torrens, H. S., and J. A. Cooper. "Uncurated Curators: No. 1, George Fleming Richardson (1796–1848)—Man of Letters, Lecturer and Geological Curator." *Geological Curator* 4, no. 5, issue 2 (1985).

Torrens, Hugh. "Mary Anning (1799–1847) of Lyme: 'The Greatest Fossilist the World Ever Knew.'" *British Journal for the History of Science* 28 (1995).

———. "Politics and Paleontology: Richard Owen and the Invention of Dinosaurs." In M. K. Brett-Surman, ed., *The Complete Dinosaur*. Bloomington, IN: Indiana University Press, 1997.

Tuckwell, William. *Reminiscences of Oxford*. London: Cassell, 1901.

Van Riper, A. Bowdoin. *Men Among the Mammoth: Victorian Science and the Discovery of Prehistory*. Chicago: University of Chicago Press, 1993.

Wallace, David Rains. *Beasts of Eden: Walking Whales, Dawn Horses, and Other Enigmas of Mammal Evolution*. Berkeley, CA: University of California Press, 2004.

Wedderburn, Alexander, and Edward Cook, eds. *The Works of John Ruskin*. Vol. 36. London: Allen, 1909.

Westfall, Richard. *Never at Rest: A Biography of Isaac Newton*. New York: Cambridge University Press, 1983.

Whewell, William. *Astronomy and General Physics. Considered with Regard to Natural Theology*. Treatise 3 of *The Bridgewater Treatises on the Power, Wisdom, and Goodness of God as Manifested in the Creation*. London: Pickering, 1833.

White, Matthew. *Atrocities: The 100 Deadliest Episodes in Human History*. New York: Norton, 2011.

Williams, David B. "Benchmarks: December 31, 1853: Dinner in a Dinosaur." *Earth*, June 20, 2016.

Williams, Paige. *The Dinosaur Artist: Obsession, Betrayal and the Quest for Earth's Ultimate Trophy*. New York: Hachette, 2018.

Wilson, A. N. *God's Funeral: A Biography of Faith and Doubt in Western Civilization*. New York: Ballantine, 1999.

———. *The Victorians*. New York: Norton, 2004.

Witton, Mark, and Elinor Michel. *The Art and Science of the Crystal Palace Dinosaurs*. Ramsbury, UK: Crowood, 2022.

Wootton, David. *The Invention of Science: A New History of the Scientific Revolution*. New York: Harper, 2015.

Worster, Donald. *Nature's Economy: A History of Ecological Ideas*. New York: Cambridge University Press, 1994.

Yong, Ed. *An Immense World: How Animal Senses Reveal the Hidden Realms Around Us*. New York: Random House, 2022.

Young, George. "Account of a Singular Fossil Skeleton, Discovered at Whitby, in February 1819." *Memoirs of the Wernerian Natural History Society*. Vol. 3, *For the Years 1817-18-19-20*. Edinburgh: Constable, 1821.

Zimmer, Carl. *Evolution: The Triumph of an Idea*. New York: Harper, 2001.

Zimmerman, Virginia. *Excavating Victorians*. Albany, NY: SUNY Press, 2007.

Image Credits

Index